To Tina Beef

with my very best wishes

Patrick Geddes's Intellectual Origins

Murdo Macdonald

EDINBURGH
University Press

Edinburgh University Press is one of the leading university presses in the UK. We publish academic books and journals in our selected subject areas across the humanities and social sciences, combining cutting-edge scholarship with high editorial and production values to produce academic works of lasting importance. For more information visit our website: edinburghuniversitypress.com

Edinburgh University Press Ltd
The Tun – Holyrood Road
12 (2f) Jackson's Entry
Edinburgh EH8 8PJ

Typeset in 10.5/13pt Sabon by
Servis Filmsetting Ltd, Stockport, Cheshire
and printed and bound in Great Britain

A CIP record for this book is available from the British Library

ISBN 978 1 4744 5407 0 (hardback)
ISBN 978 1 4744 5408 7 (paperback)
ISBN 978 1 4744 5409 4 (webready PDF)
ISBN 978 1 4744 5410 0 (epub)

Contents

Figures

Preface

Patrick Geddes was one of the key thinkers of the late nineteenth and early twentieth centuries. His integrated ecological and cultural vision had international impact. Yet despite the fact that his Outlook Tower and Ramsay Garden are prominent and intriguing features of the Edinburgh skyline, the distinctively Scottish intellectual context of his thinking has not been understood. This book places Geddes with respect to his own intellectual background, a background which had interdisciplinarity at its heart. It explores the relevance of that generalist view for his achievements local, national and international.

Note on Abbreviations
(i) NLS: National Library of Scotland;
(ii) SUA: Strathclyde University Archives.
Other collection details are given in full.

Acknowledgements

Bert Barrott, Halla Beloff, Michael Bolik, Chris Bonfatti, Anna-Maria von Bonsdorff, Tom Bower, Lise Bratton, Caroline Brown, Veronica Burbridge, Abigail Burnyeat, Grant Buttars, Claire Button, Richard Carr, Hugh Cheape, Neil Christie, Russell Clegg, Cairns Craig, Elizabeth Cumming, Tricia Cusack, Elspeth Davie, George Davie, Giancarlo De Carlo, Alec Finlay, Jan Magnus Fladmark, Frances Fowle, Bashabi Fraser, Howard Gaskill, Claire Geddes, Colin Geddes, Marion Geddes, James Gray, Murray Grigor, Richard Gunn, Wendy Gunn, Ysanne Holt, Tom Hubbard, Nicola Ireland, Robert Alan Jamieson, Matthew Jarron, Alan Johnston, James Kelman, Anjam Khursheed, Sabine Kraus, Peter Kravitz, Gauti Kristmannsson, Alex Law, Julie Lawson, Sofia Leonard, Lesley Lindsay, Michael Lynch, Archie MacAlister, Euan McArthur, Donald MacCormick, Mairi McFadyen, Elaine MacGillivray, Jim McGrath, Jean McGuinness, Alastair McIntosh, Charles McKean, Kenneth Maclean, Will Maclean, Finlay MacLeod, Duncan Macmillan, Jan Merchant, John Morrison, Kenny Munro, John Stuart Murray, Riitta Ojanperä, Kiyoshi Okutsu, Lindsay Paterson, David Patterson, Mary Pickering, Geraldine Prince, Fergus Purdie, John Purser, Graeme Purves, Tessa Ransford, Jennie Renton, Gavin Renwick, Siân Reynolds, Alan Riach, Una Richards, Lou Rosenburg, Noboru Sadakata, Tom Schuller, Boris Semeonoff, Michael Shaw, Norman Shaw, Mike Small, Joanna Soden, Fiona Stafford, Ian Stephen, Walter Stephen, Margaret Stewart, Domhnall Uilleam Stiùbhart, Annie V. F. Storr, Will Storrer, Ninian Stuart, Robert Tavernor, Belinda Thompson, John Tuckwell, Donald Urquhart, Lorna J. Waite, Arthur Watson, Volker Welter, Chris Whatley, Pat Whatley, Barrie Wilson, Peter Wilson, George Wyllie, Clara Young.

My debt is considerable to the work of Geddes's biographers Philip

Boardman, Paddy Kitchen, Philip Mairet and Helen Meller, and to the work of the Sir Patrick Geddes Memorial Trust.

The following libraries and their associated archives and special collections have been invaluable: Dundee Public Libraries (Lamb Collection), Edinburgh Central Library (Edinburgh Room), National Library of Scotland, University of Chicago, University of Dundee, University of Edinburgh, University of Strathclyde, University of St Andrews. My thanks to every librarian and archivist who made my task easier. Particular thanks are due to the University of Dundee Archive Services for providing an image of Geddes's sketch of his valley section diagram, and for giving me permission to reproduce it.

The basis of this book was laid in 2001 and 2002 during a period of research leave from the Department of Modern History and the Department of Fine Art (Duncan of Jordanstone College of Art and Design) at the University of Dundee. The funder was the Arts and Humanities Research Board. Thanks are due to all the journal and book editors who enabled me to refine my ideas (see Appendix 2 for a list of those publications). The text has improved with time.

This book is dedicated to the memory of my teacher George Davie, author of *The Democratic Intellect* and *The Crisis of the Democratic Intellect*.

Murdo Macdonald
Edinburgh and Stornoway
21 June 2019

1

Patrick Geddes and the Scottish Intellectual Tradition

A GENERALIST AMONG GENERALISTS

PATRICK GEDDES WAS ONE of the most remarkable thinkers of the late nineteenth and early twentieth centuries. He was an ecologist before that discipline had even been defined. He was a geographer of note. He was a highly original theorist of cities. He was one of the founders of sociology. He was an advocate of the importance of the arts to everyday life. He was a community activist. He was an internationalist. He was a cultural revivalist. He was a college designer. He was a publisher. He made gardens from city gap sites. He is often referred to as 'the father of town planning'. He was professor of botany at University College Dundee. He was professor of sociology and civics at the University of Bombay. He was appointed a fellow of the Royal Society of Edinburgh while still in his twenties. In short his achievements were substantial, interdisciplinary and international. No one has better characterised Geddes than his friend Rabindranath Tagore, who wrote that he had 'the precision of the scientist and the vision of a prophet; and at the same time the power of an artist to make his ideas visible'.[1]

The purpose of this book is to give context to Geddes's achievement with respect to his own place, namely Scotland. His Scottish background was not simply a question of geography, languages and history, it was a question of intellectual tradition, for Geddes was born into an intellectual culture which had interdisciplinary thinking at its heart. The classic account of that tradition has been given by the Edinburgh philosopher George Davie in his book *The Democratic Intellect*, published in 1961, and in its sequel *The Crisis of the Democratic Intellect*, published in 1986.[2] Davie notes Geddes as the inheritor of an intellectual approach that emphasised the importance of interdisciplinary thinking. In *The*

Democratic Intellect he explores how, in the old Scottish curriculum, study of the core discipline of philosophy enabled students to explore and compare disciplines in relation to one another. Writing of the 1830s, Davie notes that 'philosophy acquired a commanding position in the higher educational system, and, interestingly enough, the other academic subjects responded to this unique situation, not by envying philosophy, but by themselves becoming more philosophical'.[3] The essence was that the technical aspects of subjects – whether mathematics or classics – were balanced by philosophical and historical dimensions. Davie cites Geddes's teaching of biology in the late nineteenth century as one of the last examples of that consciously philosophical interpretation of the sciences in Scottish education.[4] He also writes of James Clerk Maxwell endorsing the value of philosophy for the scientist.[5] Maxwell was until his death in 1879 the science editor of that interdisciplinary compendium of arts and sciences, the ninth edition of *The Encyclopaedia Britannica*.[6] In due course Geddes himself would be a contributor. George Davie mentions two other Scottish academics linked to Geddes: the literary critic and historian David Masson (another *Britannica* contributor) and the educationist Simon Somerville Laurie. Like Maxwell they were of the generation before Geddes but he knew them both. And like Geddes they were both interdisciplinary thinkers of international vision. Aberdeen-born Masson was professor of English literature at University College London and then at the University of Edinburgh. His major research contribution was a monumental six-volume *Life of Milton*. Laurie was professor of education at Edinburgh, where he pioneered the development of teacher training based on developmental psychology. Laurie and Masson came together in their shared interest in the history of interdisciplinary thinking, and it is there that they link directly to Geddes for all three shared an interest in one of the early modern theorists of interdisciplinary knowledge, the seventeenth-century Czech educator, Jan Amos Comenius. We call such thinking 'interdisciplinarity' or 'generalism' or 'the democratic intellect'.[7] Comenius called it 'pansophism'. In the seventeenth century he put the rationale for such thinking like this: 'He deprives himself of light, of hand and regulation, who pushes away from him any shred of the knowable'.[8] That is to say, you must take note of the diversity of what can be known and the different ways of knowing it; if you do not, you will lack insight, you will not know how to act and you will lack control. The quotation appears in Masson's *Life of Milton*, so Geddes may have read those very words. He would certainly have read Laurie's book on Comenius published in 1881 and reprinted numerous times. In that book Laurie was the first scholar to bring Comenius's work fully to the attention of English-speaking educationists in the nineteenth century.[9] He was not simply an advocate of Comenius's interdisciplinary thought; his own educational theories drew on the stages of childhood psychological development that Comenius suggested.[10] Laurie also notes that 'Comenius's inspiring motive, like that of all leading educationalists, was social regeneration'.[11] He could have been writing of Geddes, who makes clear his view of Comenius's generalist approach in *Cities in Evolution*, published in 1915: 'among the educators of history there are few more significant and perhaps none at this moment more vividly modern' than Comenius. And he notes that Comenius emphasises 'the twofold needs of progress of sciences and humanities together'.[12] Reading that passage without knowledge of Geddes's Scottish intellectual background one might think it simply an indication of his international awareness. But reading it in knowledge of Geddes's links to Laurie and Masson gives one an appreciation of how that Scottish background enabled him to engage with interdisciplinary thinking wherever he found it. It is no surprise to discover that in 1896 Geddes invited Laurie to be an advisor to his interdisciplinary Edinburgh summer meeting, a pioneering example of the higher education summer schools we now take for granted.[13]

DISCIPLINES AND INTERDISCIPLINARITY

Geddes understood the value of contact between disciplines. His close contemporary, John Burnet of the University of St Andrews, put it like this: 'every department of knowledge has its universal side, the side on which it comes into touch with every other'.[14] Burnet and Geddes had no particular link to one another but the point is that they were both part of the same intellectual tradition. Burnet was a professor of classics, Geddes was a professor of botany, but both understood that disciplines depend for their vitality on interdisciplinary awareness. In 1917 Burnet wrote in defence of that generalist position that 'it is extremely desirable that students of the Humanities should know something of Science, and that students of Science should know something of the Humanities'.[15]

As a pioneer of sociology, of planning and of ecology, Geddes would have been keenly aware that academic disciplines emerge from the interaction of earlier formulations of study. It is an irony that as disciplines develop into recognisable areas of endeavour their interdisciplinary origins may no longer be seen as relevant, and the significance of their relationships to other disciplines may no longer be perceived. Indeed, it is often in the professional interests of the practitioners of a new discipline to demarcate it clearly from other disciplines. Geddes's 'diluted legacy' in planning, geography and sociology has been noted by B. T. Robson,[16] who comments that 'the bones, not the spirit' of Geddes's thinking were taken up in those disciplines.[17] That lost spirit was, in large part, Geddes's interdisciplinarity.

My task is to explore Geddes's activities in such a way as to illuminate that interdisciplinary richness rather than to regard it as an inconvenient distraction from a more specialised path. From that perspective the earliest commentaries on Geddes's work are among the most useful. In 1914 his colleague Victor Branford published *Interpretations and Forecasts*, a book which draws heavily on Geddes's interdisciplinary approach.[18] Branford was no mean thinker in his own right and the purpose of his book was to reflect on the value of the nascent discipline of sociology which he was, with Geddes, at that time helping to bring into being.[19] But the first book to attempt a full account of Geddes's approach was *The Interpreter: Geddes, the Man and his Gospel* by Amelia Defries. It was published in London in 1927 and in New York the following year.[20] It remains a key source of Geddes's thinking, both verbal and visual. The book was published with Geddes's approval and involvement, and as such it is invaluable. Yet neither Branford nor Defries illuminates Geddes's Scottish intellectual background. For that one must turn to one of his Scottish peers, a student of Simon Somerville Laurie, the educationist Stewart Alan Robertson.[21] His essay collection *A Moray Loon* was published in 1933 and with respect to understanding Geddes it is important for two reasons. First of all, it contains an insightful essay about Geddes himself. Second, the book as a whole is an expression of that internationally inclined, culturally aware, Scottish generalist tradition of which both Robertson and Geddes were part.[22] Most commentators on Geddes have simply not been conscious of that intellectual tradition, even though it is fundamental to his thinking.[23]

Two industrialised wars fostered disciplinary specialisation in the twentieth century, and the Second World War was a watershed for how Geddes was considered. Despite the best efforts of his self-styled disciple, the American Lewis Mumford, after that war Geddes's interdisciplinary approach was seen as an eccentric quality, not of importance in its own right. Yet just as that aspect of Geddes's thinking was fading from public consciousness, C. P. Snow was exploring his 'two cultures' debate and that debate became part of public consciousness as though there had been no previous consideration of the relationships between arts and sciences in the twentieth century. As the confident high modernism of the immediate post-Second World War period dissipated, the broad view that Geddes represented again became of interest. Nineteen sixty-one saw the publication of George Davie's *The Democratic Intellect*, and in 1966 Davie's friend the poet and essayist Hugh MacDiarmid wrote of both Davie and Geddes in his book, *The Company I've Kept*. Intellectual generalism was again on the agenda and not only in Scotland. In *Mind and Nature*, published in 1979, in a most Geddes-like way the English anthropologist Gregory Bateson summarised the argument for generalism as follows: 'break the pattern which connects the items of learning and you necessarily destroy all quality'.[24] But by that time Geddes's relevance to the debate was little noted even in Scotland and his reputation was seen primarily in terms of his role as a 'father of town planning' with, to repeat Robson's point, the spirit of his thinking for the most part lost.

VISUAL THINKING AND INTERDISCIPLINARITY

The ability to think visually is another important aspect of Geddes. The linkage between taking a broad view of knowledge and thinking visually is interesting. It is not hard to see why that linkage should exist for there is a holism in a visual approach that facilitates the multiple connections which are required by a generalist view of knowledge. George Davie helps one to understand that in the context of Scottish thought by making the link explicit in his discussion of eighteenth-century mathematicians. Just as had Newton a century earlier in England, Scottish Enlightenment thinkers advocated a geometrical rather than an algebraic approach to mathematics. A key figure was Robert Simson, whose influential edition of Euclid was first published in 1756 and went into numerous editions, both under Simson's name and in an expanded version edited by his pupil John Playfair. From the perspective of generalism, a crucial quality of geometry is that it seems to be a purely mental activity but is at the same time real to the senses: it thus crosses modalities of thought and perception, and facilitates interdisciplinary understanding. For Davie, 'geometry, at least if taught in the Greek spirit, connected up with the other principal disciplines in a way that the more specialised techniques of algebra did not', and he continues by noting that 'in view of the broad, general approach, characteristic of Scottish education, this predilection for Greek geometry was both reasonable and natural'.[25] It is illuminating that Geddes remembered his Perth schoolteacher's emphasis on geometry as 'the greatest educational influence of all'.[26]

Further indication of the significance of that commitment to the visual and its concomitant generalism in Scottish intellectual culture can be inferred from the fact that Geddes wrote the entry on 'morphology' for the ninth edition of the *Encyclopaedia Britannica*, while James Clerk Maxwell had written the entry on 'diagrams' a few years earlier.[27] Geddes's colleague at University College Dundee, the biologist D'Arcy Wentworth Thompson, also contributed to the ninth edition and in due course wrote that classic of visual thinking about biology, *On Growth and Form*, published in 1917, which followed two years after Geddes's strongly visually informed book, *Cities in Evolution*. An understanding of Geddes must, therefore, take note of the Scottish generalist tradition for its visual as well as its interdisciplinary aspects.[28] What flows from that is a further strengthening of the appreciation of Geddes's Scottish intellectual background as psychologically central to his wider interdisciplinary achievement.

PRESBYTERIANISM AND INTERDISCIPLINARITY

It may be surprising from a present-day perspective to draw attention to the dominant religious current of nineteenth-century Scotland, namely Presbyterian Christianity, as contributing to Geddes's interdisciplinary mindset. One can get some notion of why that might be important from the historian Norman Stone:

> In some respects, the Scottish Enlightenment, in the eighteenth century, had been an anticipation of later developments in Vienna: the same desire to systematise, to overthrow outworn structures, to rationalize. The secularisation of the Calvinist mind, and the secularization of the Jews, gave early twentieth-century intellectual life its characteristic stamp.[29]

That notion of 'the Calvinist mind' sheds light on the intellectual significance of that particular aspect of Christianity for Geddes. Geddes's father was an elder of the Free Church of Scotland, his mother was deeply religious and even hoped that Patrick might become a minister. That has led some commentators to consider religion as something that Geddes had to get away from in favour of a scientific approach, but that is to misunderstand the Scottish Presbyterian churches of the second half of the nineteenth century. Geddes's religious background has to be considered from two perspectives: first of all from the perspective of 'the Calvinist mind' and secondly with respect to the specific dynamics of the Calvinist organisation into which he was born in 1854, namely the Free Church of Scotland. From the first perspective one must note the democratic basis of Presbyterianism. That particular form of church government, which originated in the thinking of the Frenchman John

Calvin and was developed by his Scottish student John Knox depends, at least in principle, on the wishes of the members who constitute the Church. The minister is appointed by the congregation. There are no bishops, and a new temporary leader of the Church, the moderator, is elected every year at a general assembly. The name 'Presbyterianism' itself refers to the elders (Greek *presbyteros*: elder) who are elected from within congregations to represent them. The key to this structure is that individual views (at least in principle) have equal weight and hierarchy is minimal. That is why Presbyterianism is prone to schism: disagreement can lead to the formation of new churches which have an equal claim to Presbyterianism. By the same token, it is prone to reunification. From the point of view of the status of different forms of knowledge, the secularisation of such a structure leads to a generalist view because within such a structure all knowledges, like the views of all church members, have an equal claim to respect. Similarly, a Presbyterian structure can be thought of as providing an example for non-hierarchical forms of political organisation. Geddes's thinking exemplifies both those secular transformations: intellectual generalism on the one hand and political anarchism (of which more in chapters three and five) on the other.

The particular branch of Presbyterianism into which Geddes was born was the Free Church of Scotland. That Church had been founded in 1843, close to eleven years before Geddes's birth, when about one-third of the congregations of the Church of Scotland seceded from the main body. Adherents of the new Church included many of Scotland's most significant thinkers. The key issue of that secession was the refusal to accept that a London parliament, or indeed any parliament or any monarch, had the right to dictate on matters relating to the appointment of ministers of religion. That is to say, it was a rejection of a hierarchy of political authority on the basis that such a hierarchy could not, in principle, operate with reference to a Presbyterian church. The Free Church was thus from its foundation onwards a site of active and continuing debate about matters of principle and how they relate to practice.

From its inception the Free Church was strongly associated with academics and artists, and that has clear relevance to the understanding of Geddes as a thinker. For example, among its founders was the physicist Sir David Brewster. Brewster had been a key figure in the establishment of the British Association for the Advancement of Science in 1831. Another founder member of the Free Church was the painter David Octavius Hill, better known today for his pioneering role with Robert Adamson in the development of portrait and social documentary photography. Hill began his collaboration with Adamson at Brewster's suggestion in order to help record the first general assembly of the Free Church of Scotland. It was thanks to Brewster's friendship with the inventor of the calotype process, the Englishman William Henry Fox Talbot, that Hill and Adamson adopted that most appropriate photographic method. Since 1830, Hill had also been the very active secretary of the first body to properly represent the interests of professional artists in Scotland, namely the Scottish Academy (from 1838 the Royal Scottish Academy). The founder of the Free Church, Thomas Chalmers, was the Scottish Academy's first chaplain, serving from 1830 through the Disruption of 1843 (the moment of creation of the Free Church) until his death in 1847. Thus those who helped to bring the Free Church into being included those who had helped to structure and support major organisations concerned with scientific research on the one hand and the making of art on the other. That conjunction of morality, spirituality, science and art set the agenda for Geddes.

Other founders of the Free Church included the geologist and journalist Hugh Miller, Thomas Guthrie, originator of the ragged school movement, and the Greek scholar and advocate of Celtic studies, John Stuart Blackie. All three would influence Geddes. A Free Church figure of a younger generation, but still a decade or so older than Geddes, was William Robertson Smith. Robertson Smith's social anthropological ideas were to influence not only his fellow Scot James Frazer (also Free Church in background), but Geddes's close contemporaries W. R. Lethaby and Sigmund Freud. The link between Freud and thinkers in a Calvinist tradition may seem somewhat surprising, but it resonates with the comment quoted above that the 'secularization of the Calvinist mind, and the secularization of the Jews, gave early twentieth-century intellectual life its characteristic stamp'. It is less surprising when one notes the Calvinist background of the most mythologically inclined of Freud's followers, C. G. Jung; indeed, a Calvinist environment seems to have been a useful starting point from which to develop thinking about mythology in the latter half of the nineteenth century.

Geddes's background in the Free Church thus provided him with a training in comparative thinking. In due course it would provide his first insight into the ironies of cultural politics: William Robertson Smith's work on the status of the Bible as history had both helped to develop the emerging science of social anthropology, and had also engendered a major debate within the Free Church with respect to how the Bible should be approached from an academic perspective. The outcome of the debate was that, in 1881, Smith narrowly lost his position as professor in the Free Church College at Aberdeen. Yet soon after he was appointed overall editor of the ninth edition of the *Encyclopaedia Britannica*. According to the entry on him in a later edition of that work, the eleventh, he also gained widespread popularity as a result of his views. As a young man Geddes was well aware of the intricacies and ironies of the Robertson Smith affair, and he even used analogy with Robertson Smith's unfettered thinking to comment – with some justification – on the undermining of his own academic prospects (of which more in chapter four).

Thus, what one finds in the Free Church during the period of Geddes's development as a thinker is an institution which transcends any stereotype of Presbyterian rigidity of view. Indeed, one can think of it as an institution which facilitated a network of wide-ranging debates on issues ranging from the status of legendary sources as history through the relationship of spiritual to temporal forms of government to the integration of evolutionary theory with Christian faith.[30] Geddes retained strong links with thinkers of Free Church origin throughout his life; indeed, one of them, the Reverend John Kelman, wrote the first detailed account of Geddes's Outlook Tower in Edinburgh.[31]

ART AND SCIENCE

The interdisciplinary current of Scottish intellectual culture in the nineteenth century also gives insight into Geddes's support for the visual arts. I will consider that in more detail from chapter three onwards, but I give a sense of it here from his final lecture to University College Dundee, given in 1919 before he took up his post as professor of sociology and civics at the University of Bombay. Geddes makes clear that he considers the visual arts complementing the sciences to be of fundamental educational importance: 'We need to give everyone the outlook of the artist, who begins with the art of seeing – and then in time we shall follow him into the seeing of art, even the creating of it.'[32] That lecture also includes one of Geddes's most memorable statements, 'By leaves we live', a phrase that has in more recent times become an ecological motto.[33] As a synopsis of why interdisciplinary thinking is important, that Dundee lecture stands the test of time a century later. I return to it in my concluding chapter.

2

Geography, History and Place

BALLATER AND PERTH

COMPLEMENTING GEDDES'S INTELLECTUAL BACKGROUND was the geography of Scotland. He was born in the Highland village of Ballater in Aberdeenshire in 1854. The village lies on the river Dee just before it leaves the Cairngorm massif. To the south is the mountain ridge of Lochnagar. To the east is farmland stretching forty miles to the coast at Aberdeen. So young Patrick's earliest awareness would have been of hills, human settlement and river in immediate relationship. Those key elements of human geography were to remain at the heart of his thinking for the rest of his life. The relationship of mountain watershed to agricultural plain, and the diverse habitats of plants, animals and human beings that such relationship provides, became further emphasised for Geddes when he was three years old. In 1857 his family moved fifty miles south to the city of Perth. His new home was a cottage on a hillside overlooking the city. Between Geddes's home and the main body of Perth ran the river Tay. In the distance was the central Highland massif. Behind his house rose the tower-topped viewpoint of Kinnoull Hill.

Thus Geddes's appreciation of the relations between city and region had its foundation in his earliest childhood sense of the mountains above Ballater defining the course of the river Dee and his subsequent appreciation of the position of Perth in the wide valley systems surrounding it: Strathearn, Strathtay and Strathmore. His understanding was shaped by boyhood rambles through the woods and among the crags of Kinnoull Hill, exploring in turn the ecology of the woodland, the geology of the hill, and the viewpoints over the river and the city. As he wrote himself, 'it must have been in the climbings and the ramblings over this fine valley landscape . . . that I got the feeling of the valley section'.[1] That 'valley section' became Geddes's key tool for geographical analysis in later years. As he developed the valley section idea, his childhood experience would be complemented by the thinking of his older contemporary, the French geographer Élisée Reclus, whose *Histoire d'un Ruisseau* traces a watercourse from source to sea, exploring how human settlement is shaped by the availability of that water as it descends. With that visual tool, Geddes made sense of basic human activities with respect to their appropriate place in the landscape.

Reference back to the river Tay was a constant of Geddes's life, always a source of illumination by analogy. When he was involved in studies in India, the Ganges – despite the difference in scale – reminded him of 'his own childhood river, the Tay, "which will always be for me my main impulse of the life-stream and of the cosmos" '.[2] The Tay thus became for Geddes a way of modelling the essential characteristics of a river region as a geographically, historically, politically and spiritually significant entity. Stewart Robertson notes that when Geddes spoke 'on the interrelations of geography and history . . . he illustrated his thesis by reference to the Thames, with its English capital city, London, and its sacred place of coronation, Westminster, paralleled by the Tay with its Scottish capital, Perth, and its sacred place of coronation, Scone'.[3]

Perth provided the young Patrick with a rich source for developing his thinking. Despite being strongly defined by high ground and rivers, the city is not cut off in any direction. It is a natural hub. Both historically and prehistorically, the surrounding region has been culturally significant. In Geddes's time Perth was noted for its textile trade. In the late eighteenth and early nineteenth centuries it was noted for its booksellers and publishers. And Geddes's childhood and youth coincided with the growth of Victorian museum culture, with Perth developing collections of distinction. Those included that of the Perthshire Society for Natural Science, founded in 1867. Geddes would later write the introduction to a book about one of the most notable of those Perthshire naturalists, Charles Macintosh, who came from Inver near Dunkeld.[4]

The Tay at Perth has been historically documented since Roman times. Agricola is reported to have greeted the Tay with the words '*ecce Tiber*', 'behold the Tiber'. Although not a permanently settled Roman area, the hinterland of Perth nevertheless has significant remains from that time, including the roads and communication stations along the Gask ridge between Perth and the strongpoint of Drummond Castle. But one is also drawn into an altogether older landscape of prehistoric standing stones, of which Geddes would have been keenly aware. Even on foot, the young Patrick would explore up to fourteen miles from his home, which gives some idea of his engagement with the region.[5] Perth itself was also rich with conflicting and complementary ideas for a curious youth. In Perth in 1559 John Knox had preached his sermon against idolatry, giving impetus to the unfortunate wave of iconoclasm which characterised the Scottish Reformation. In Perth, cultural and linguistic difference between Highland and Lowland Scotland were part of everyday life. In Perth, Jacobite and Hanoverian soldiers had come and gone in the eighteenth century. In Perth, Sir Walter Scott set one of his great novels.

In his *Talks from the Outlook Tower*, Geddes evoked the geography and culture of Perth thus:

> From the corner of the hill nearest the city you could look down upon it, lying beautiful between its north and south 'Inches'. These were two large old parks, each (as the Gaelic name means) islanded between river and mill-streams, ascribed to Roman origin. Perth is still something of 'the Fair City' its folk have long called it, and from the rock-ridge across the river you look down into it, almost as on a map, say rather a relief-model in perspective. Below our hillside home the river-mist would sometimes lie over and conceal it, in a long grey-white lake, with only the spires rising through . . . And in scanning these two aspects of my home-city again in memory, I realise that there were the best of preparations a town-planner could desire; at ordinary times the precise observation of the city in detail; yet at others the discernment of its old ideals, emergent above the mist of nature and the smoke of its working life.[6]

THE BLACK WATCH

Further Scottish cultural and historical influences came from a military source. Perth was the headquarters of the regiment of the British army known since its inception in 1740 as the Black Watch, and Geddes's father, Alexander, had his career as a soldier in that regiment. The regiment had very strong Highland roots and in the eighteenth century had been composed of Gaelic speakers only. In the first half of the nineteenth century a Gaelic speaker such as Geddes's father would have found himself very much at home.[7] Indeed, Patrick's daughter Norah, in her memoir of her father, notes that Alexander Geddes served in the Black Watch at a time 'to quote a Northern newspaper ". . . when its officers were Gaels who rose from the ranks to command that proud old regiment" '.[8] Alexander Geddes had joined up as a drummer in 1826 at the age of fifteen, and he left with the rank of acting sergeant-major in 1851, but continued for another seven years as sergeant-major of the reserve battalion.[9] His career with the Black Watch affected his family in ways that can be seen from points of view both local and international. For example, Patrick's elder sister was born in Corfu when the family was stationed there and Alexander Geddes spent more than half his service career abroad. After his discharge in 1851 he took his family to live at Ballater in Aberdeenshire, where Patrick was born.

When, in 1857, the family moved to Perth it may be that Alexander Geddes's presence as a reservist was required because of the Black Watch's posting to India.[10] The main body of the regiment was engaged in the suppression of an early aspect of the Indian struggle for independence from the British Empire, what was known as the Indian Mutiny, so India would have been an everyday topic of conversation in the Perth of Geddes's childhood. The participation of troops from indigenous Indian regiments in the mutiny was an expression of the cultural tensions of Empire, and it is instructive to note that one can consider the history of the Black Watch from that perspective also. First mustered as a regiment in 1740, the Black Watch had in 1743 been involved in a mutiny engendered by cross-cultural suspicion. Indeed, a significant number of those who were involved in the Black Watch Mutiny were monoglot Gaelic speakers. Despite the relatively minor nature of the Black Watch Mutiny, it resulted in executions and punishment postings to other regiments. In 1745 the regiment was held back in Kent rather than risk divided loyalties during the Jacobite war. It is an irony that the harsh treatment meted out to the Black Watch mutineers may have influenced the decision of a number of Highland clans to support that Jacobite uprising.[11] So there are indigenous cultural analogies between the soldiers of the Black Watch and the soldiers they were sent to suppress in India. Thus a sense of the interplay of imperial power and indigenous cultures was woven into Geddes's background. That gives context to his later understanding of the dynamics of cultural revivals not only in Scotland but throughout the world, not least in India.

There is a further link to the Black Watch that should be noted here. During the early period of the Black Watch's history the regiment's chaplain was Adam Ferguson, subsequently professor of moral philosophy at the University of Edinburgh and a figure whose study of civil society Geddes would value. Geddes helped to found the Sociological Society in the early twentieth century, but, as Donald MacRae has noted, 'sociology began with Ferguson' in the eighteenth.[12] It is no coincidence that a pioneer of sociology should have served in the Black Watch during the period of the Jacobite war of 1745–6. What better insight could one have into the complexities of society than to be a Gaelic-speaking

Highlander in a regiment sworn to serve the interests of those intent
on the cultural destruction of Highland Gaelic culture? A remarkable
response to this fraught situation was James Macpherson's re-creation
of Gaelic legend in *The Poems of Ossian* published from 1760 to 1773.
It is instructive to note that Ferguson, a native Gaelic speaker, was
a defender of Macpherson's work.[13] In 1896 Patrick Geddes would
publish an edition of *Ossian* as one of his Celtic Library projects (see
chapter seven).[14]

By the time it had become part of Patrick Geddes's cultural background, the Black Watch was well travelled and had a formidable
reputation, adding to its battle honours, among others, Ticonderoga,
Alexandria, Salamanca, Quatre Bras and Waterloo. However, the period
of his father's service coincided with the long peace after the defeat of
Napoleon and before the outbreak of the Crimean War. That long
period of peace is of interest. The Black Watch spent most of it either in
Scotland or on station in various parts of the Mediterranean, and one
wonders how much his family's reminiscences of the Ionian isles brought
life to Geddes's later classical studies. Intriguing in view of his later use
of pageantry as an educational tool is a comment with respect to that
period that 'as always in peace-time, considerable attention was paid to
uniform detail. The first half of the nineteenth century was the era when
military uniforms were at their most gorgeous, and correspondingly at
their most impractical.'[15]

It was during that extended post-Waterloo period of peace that the
long, ornamental sporran became popular. During the Napoleonic wars,
sporrans, if worn at all, were for practical reasons quite modest in size,
but by the time of the Crimean war they extended well past the hem
of the kilt.[16] The sporrans that the young Patrick Geddes is pictured
wearing at the ages of ten and fifteen are very much part of that enter-
tainingly excessive school of sporran development.[17] Military Highland
dress can be seen as a developing set of clothing and accessories which
has aspects both of practicality and display. It is not surprising that
during the long period of peace display dominated practicality; indeed,
the years around 1844 have been called the time of 'the peak of sartorial
magnificence'.[18] In the case of the Highland regiments the whole process
was boosted by Sir Walter Scott's stage management of George IV's visit
to Scotland in 1822 but it is important not to regard Highland dress
as being merely the construction of that novelist's (or a Black Watch
quartermaster's) imagination. The type of Highland dress often attributed to Scott's influence was fully established well before he penned his
first word about the Highlands. Indeed, the first major work of Scottish

painting depicting what we think of as Highland dress dates from the seventeenth century.[19] It is an irony that some of the earliest widely available images of Highland dress were depictions of the Black Watch mutineers of 1743.

The Highland dress of the Black Watch was closely related to the development of the military pipe band. The Highland bagpipe – part of a family of Indo-European reed pipes which extends from Scotland to the Himalaya – was fundamental to the culture of the Black Watch, and Geddes as a child thus had an inside track for the appreciation of both pageantry and pipe music, subjects which he was to explore extensively in later life in his masques and mural schemes. Such pageantry and music were thus for Geddes parts of a living tradition. The long sporran may have been a military fashion innovation, but it was a fashion innovation that related to traditional culture. That approach to tradition as living culture can be noted with respect to Geddes's later role as the central figure of the Celtic revival in Scotland. He valued the past but he was unafraid of change. Cultural authenticity was for Geddes part of a lived experience, not an archaeological discovery. In a section on tartan in an unpublished paper on Celtic art, Geddes notes the tartan of 'the 42nd' without further comment.[20] The reference is clear to anyone familiar with the Black Watch as the 42nd Regiment of Foot; to anyone else it is obscure. What this indicates is the degree to which Geddes took his own awareness of the Black Watch for granted. Thus the Black Watch was an effective cultural mediator for the young Patrick Geddes.

GARDEN AND WORLD

While on the one hand Patrick's father linked him to the Celtic cultural background of the Black Watch, on the other he was his first practical instructor in an art that Geddes was to value throughout his life: gardening. And not just gardening in itself but gardening as the psychological heart of a wider appreciation of human interaction with the natural world. By the time he was being tutored in the beauties of geometry by Rector Miller at Perth Academy, Geddes's practical training as a gardener was well under way. In his *Talks from the Outlook Tower*, he reflects on his early gardening experience in a way that enables us to begin to understand his motto, *Vivendo discimus* – 'by living we learn'.

> I can see that my main good fortune lay before school days in a home modest enough in ordinary ways, but with a large garden; – ample fruit-bushes, apples and great old wild cherry trees; with vegetables mainly cared for by my father, and a fair variety of flowers, to which my mother was devoted. I

trotted in turns after both, and thus learned to help; as also to climb, to tame robins, to keep pets and so on.[21]

For Geddes the garden was an ordered starting point for looking further:

> A great landscape too from our hillside windows; of which the range 'from Birnam Wood to Dunsinane' was but a quarter, and even that not the finest; a landscape that stretched over city and river, plain and minor hills, to noble Highland peaks, clear-cut against the evening sky.[22]

Those peaks – in particular the sharp outline of Ben Vorlich accompanied by the ridge of Stùc a' Chroin – had become an integral part of the canon of British art during the time of Geddes's childhood, for they had been portrayed in the background of John Everett Millais's *Autumn Leaves*. That was painted in 1855–6 not far from the house into which the young Patrick would move in 1857. Geddes then shifts our perspective back to the near at hand, beginning the journey out of the garden, conveying the beauties and opportunities of 'fields, and a pool and ditch, rich in insect-life and wild flowers' behind his home leading into 'a bit of moor with wild roses and golden gorse, and in this moor a large deep quarry in a basalt-dike' and then the larger mass of Kinnoull Hill itself: 'a nobly wooded hill, with fine old fir-masses and beech glades, and lovely birch here and there between'.[23]

He then moves to the wider perspective again from 'a really glorious hill-top . . . some 700 feet above the Tay and its rich alluvial plain'. The hill extended into 'a long range of noble precipice, finer than any along the Rhine . . . with a fresh southward hill-panorama, complemental to the westerly and northern one of home'.[24]

And finally he returns us to the immediacy of geological exploration and botany: 'in the quarry there were quartz crystal masses to be found; and along the precipice and its screes of broken rock one could hunt for agates'. And crucially for the botanist-to-be: 'ferns too, in variety, could be brought back for shady nooks and corners in the garden, and rockeries built from them with the beautiful stones of quarry and cliffs'. He concludes: 'Thus I made my first botanic garden!'[25] That passage, which begins and ends with a garden and between times ranges far and wide, epitomises Geddes's ecological and regional view.[26]

Norah Geddes notes that 'As soon as he could read, his father nurtured him on the Book of Proverbs whose exhortation "Wisdom is the principal thing, therefore get wisdom and in all thy getting get understanding", his son took to heart.'[27] Later his father enhanced his school education by organising classes in both art and cabinet making to complement his written studies.[28] He then did his best to prepare Patrick for a solid career, namely in banking. As a seventeen-year-old Patrick gave the National Bank of Scotland in Perth a fair trial from September 1871 for eighteen months. He left with an excellent reference, but by then his need for higher education was pressing and, in Norah Geddes's words, his father 'let him go to study chemistry and geology in Dundee and Edinburgh'.[29]

DUNDEE, EDINBURGH AND LONDON

Geddes's student days in Dundee in 1873 and 1874 deserve consideration. It was the period during which he began to teach others, if informally. Writing many years later, his lifelong friend Robert C. Buist recalled first being inspired by Geddes when he, Buist, was a schoolboy at the High School of Dundee and Geddes was a senior student at the Dundee YMCA science laboratories.[30] That was also the time when Geddes began to make the scientific contacts in Dundee that would encourage him in due course to pursue an academic career in that city. At that time Dundee was developing its regional role as higher education provider for Perthshire and Angus but it was still a decade before the establishment of University College Dundee, the institution that would provide Geddes with academic stability as professor of botany.

Despite being intrigued by geology, in 1874 Geddes shifted focus to the study of biology. He enrolled at the University of Edinburgh but for reasons still not clear he gave up after a week. Perhaps that was because it was his first exposure to what he later described as 'necrology', that is to say the study of dead specimens, rather than his own notion of biology, which emphasised the living creature. The classification of formerly living things in a lecture hall in the heart of a built-up city may have been too much of a shock after the living glories of Kinnoull Hill. Nevertheless, Geddes had been deeply attracted to Edinburgh when he had visited a few years earlier with his father and he was to return to the university in due course as a demonstrator, and devote much of his life's work to the city. When he returned to Edinburgh, part of his contribution to the city was the provision of gardens, and teaching in those gardens. Was that a reflection of what he had felt to be lacking when he came to the city as a student? In any case, in 1874, instead of studying in Edinburgh, he became enthused with the idea of studying evolutionary theory with Thomas Huxley in London, inspired by a reading of Huxley's *Lay Sermons*. Geddes was not concerned about getting a degree, he simply wanted to study with Huxley, and that valuing of apprenticeship to a particular teacher is characteristic of him. However, despite his high hopes London was no more congenial to him than Edinburgh. Before he could join Huxley's zoological laboratory he was required to take a year of more chemistry, physics and geology, and he could get no credit for his studies in Perth and Dundee. As Boardman notes: 'If Edinburgh gave him a brief glimpse of the respectable and fossilized teaching of botany, London revealed academic officialdom in all its august stupidity.'[31] But Geddes's determination to be taught evolutionary theory by Huxley saw him through and his extended stay in London enabled him to make good use of its cultural opportunities, in particular the art galleries. During his time in London Geddes stayed with his Perthshire-born friend, Ruskin's follower James Burdon.[32] It is characteristic of Geddes's breadth of view that he found himself living with a Ruskinian while waiting to be taught by Huxley.

Huxley's teaching was important to Geddes both for its intrinsic value, and for its value as something Geddes could critique. From the first point of view Huxley was the key exponent of Darwin's ideas. Indeed, there was a memorable day in the laboratory when Geddes found Darwin looking over his shoulder to get a glimpse of what he was studying under his microscope.[33] Huxley was also keenly interested in how species exist together in a geographical region and that helped to germinate in Geddes the seeds of the science of regionalism which he would later develop through the thinking of Élisée Reclus and Frédéric Le Play.[34] But from the second point of view Geddes found himself critical of Huxley's competition-based interpretation of Darwin's thinking, instead favouring a view which took co-operation and community as the driving principles of evolution. In that he was influenced by the writings of Herbert Spencer.[35] According to Philip Boardman, Huxley's criticisms of Spencer simply encouraged Geddes to re-read him.[36] At the same time Geddes found what he considered to be Huxley's too easy rejection of the positivist ideas of Auguste Comte a good reason to seek out those ideas for himself. In his introduction to the biography of the Comtean thinker John Henry Bridges, Geddes reflected on how as a student in London in 1874 he read about and pondered over Huxley's controversy with Comte's supporter Richard Congreve, so much so that he concluded that 'Huxley's criticisms were somehow missing the essential significance of the new doctrine'.[37] Subsequent contact with Congreve, Bridges and other Comteans such as Frederic Harrison and Edward Beesly reinforced that view. In due course, in 1881, it was the Comtean academic John Ingram of Trinity College Dublin who gave Geddes his first opportunity to engage in extramural teaching.[38] As Helen Meller has pointed out, the 'connection between nationalism, cultural identity and social endeavour' in Ireland made a strong impression.[39]

It is typical of Geddes's generalism that he should admire Huxley while at the same time exploring what he saw as the blind spots in his thinking.[40] But under Huxley's tutelage Geddes distinguished himself as a biologist and he left Huxley's laboratory as a qualified biological demonstrator, a skill which he employed first of all at the University of London and then at the University of Edinburgh. And – crucially – Huxley found Geddes a place with the distinguished French biologist Henri de Lacaze-Duthiers at his marine station at Roscoff in Brittany in 1878 and 1879.[41] That, and a fact-finding visit to the marine biological station in Naples, provided Geddes with the skills he needed to help set up a marine station for the University of Aberdeen at Cowie near Stonehaven in 1879.[42] Soon he would contribute to the marine station at Granton on the river Forth, established by Sir John Murray in 1885, and its successor at Millport on the firth of Clyde.

FRANCE AND MEXICO

Geddes's time with Lacaze-Duthiers at Roscoff drew him further into French culture. The Auld Alliance between Scotland and France – formalised politically in the fourteenth century – was an intellectual alliance that provided stimulus to Geddes's thinking. It had found expression again in early nineteenth-century intellectual connections. As George Davie notes, in philosophy that had been exemplified by Victor Cousin's admiration for the writings of Thomas Reid, and his debates with William Hamilton and his circle.[43] Davie also notes the influence of Scottish geometry on the influential French mathematician Michel Chasles, and Auguste Comte's employment as a translator of work by the Scottish geometer John Leslie.[44] The ideas flowed both ways: one historian of geography has noted that 'the Edinburgh–Paris axis, around which Geddes's own intellectual life revolved, had been in service since the early nineteenth century as a means of importing French political and scientific radicalism into Britain'.[45] Geddes's period with Lacaze-Duthiers enabled links in particular to the work of the sociological thinker Frédéric Le Play.[46] It was from Le Play that Geddes derived his three interacting descriptive categories of place, work and folk, which he would in due course use as the analytical basis of his regional surveys. During the winter of 1878 he attended a lecture at the Sorbonne by Le Play's follower Edmond Demolins. Siân Reynolds, in her assessment of Geddes's academic and political contacts in France, sums up the significance of Comte and Le Play for him as follows: 'what Geddes borrowed from Le Play was the need for a thorough sociological survey of a region, which became one of his key preoccupations, while in Comte he found the concept of the hierarchical structure of knowledge'.[47] Geddes's interdisciplinary background equipped him to synthesise those approaches.

Geddes's final educational experience of place as a young man was in the autumn of 1879, when he crossed the Atlantic for the first time to research for a year in Mexico. There he received the help of his elder brother, Robert, who was a bank manager in Mexico City. He was supported by the British Association and intended to collect biological specimens for Huxley, geological specimens for James Geikie and fossils for the British Museum, and indeed had some success in this. But in Mexico Geddes was to suffer an attack of temporary blindness which led to him having to take care with his eyesight for the rest of his life. That traumatic event had an unexpectedly positive consequence for it impelled him to explore his visual imagination in a different way, namely as a method of ordering knowledge. He did this in the first instance through the sense of touch, feeling the panes in his window and imitating their regular distribution in folded paper. His schemes for representing both Comte's hierarchy of the sciences and Le Play's parameters of social organisation derive from this simple folding technique, which yielded a matrix of cells. That method of disciplinary classification leading to interdisciplinary insight found its mature form in what Geddes called his notation or chart of life.[48] Philip Boardman notes that on his return journey in March 1880, 'his luggage consisted of crayfish for Huxley, fossils for the British Museum, deep-well boring samples for Geikie and graphic thinking-machines for himself'.[49]

On returning from Mexico, Geddes took up a job at Edinburgh University as demonstrator in biology, a role that gave him the opportunity to teach in the laboratory and also to lecture on biological topics. He was assistant to Professor Alexander Dickson, who become his firm friend. On medical advice he was, however, limited to two hours of microscope work per day. In his entry on his father in the *Dictionary of National Biography*, Arthur Geddes suggests that a more orthodox biologist might have found that limitation an acceptable career constraint but the unorthodox Patrick – forever wanting to find evidence for his ideas – would have found that difficult. In Arthur's view that restriction prompted his father to devote more time to his other interests: planning, sociology, geography, ecology and cultural activism. Arthur describes Geddes's temporary loss of sight as 'the supreme crisis of his intellectual life' but sees it as compensated for by the beginning of methods of diagrammatic thought on the one hand and on the other his interest in 'the founding of self-governing university halls, linked to rebuilding and the redemption of neighbourhood and community'.[50] There is no doubt that Geddes would have been an intellectually wide-ranging biologist in any event but – after a strong focus on biological research from his years with Huxley until his Edinburgh appointment – by the mid-1880s his achievements were much more diverse. That is consistent with Arthur's view that the issue with his sight moved Patrick Geddes's thinking in an even more generalist direction.

3

Arts, Crafts and Social Reform

THE EDINBURGH SOCIAL UNION

Patrick Geddes applied his interdisciplinary insight to the pressing social issues of his time. The first notable expression of that was in Edinburgh. His return to the city in 1880 marked the beginning of a period of intense activity during which his ideas of urban conservation and renewal developed. At the heart of that was a commitment to evolutionary thinking and cultural revival. His focus was the Old Town of Edinburgh, an area of the city that stretches down a spine of rock from the castle to the palace of Holyroodhouse. As its name implies, it is the site of the first habitation and expansion of the city. It exists in marked architectural and social contrast to the neoclassicism of the New Town, begun in the late eighteenth century to house Edinburgh's wealthier citizens, and in consequence drawing resources away from old Edinburgh. By the latter half of the nineteenth century the Old Town – whatever its historical interest – was characterised by dilapidated buildings and social deprivation. Thanks to Lou Rosenburg and Jim Johnson it is now possible to appreciate the intricacies of the relationship between Geddes and the city authorities in his efforts to address that situation.[1]

Geddes thought about the city as he would think about any other organism, in terms of structure, function and possibilities. Where he saw possibilities he acted. The Edinburgh Social Union was established at Geddes's instigation in 1884 as a body to facilitate such action both in terms of urban conservation and in terms of major commissions of new art. It was directly inspired by Octavia Hill's housing work in London. Hill had trained as a painter with John Ruskin, before shifting direction – with Ruskin's encouragement – to her pioneering housing work.[2] Such interlinkage of art and social reform was fundamental to the Edinburgh Social Union, which brought together financial philanthropy and volunteer activism in a structure that sought to enhance the city both through conserving its fabric and giving viability to its communities.

Geddes and his circle made Edinburgh's Old Town into a place of practical experiment in urban conservation and community renewal. His friend the French geographer Élisée Reclus wrote of these activities:

> We are told that in Edinburgh, the lovely Scottish capital, pious hands are at work ... breaking in upon its picturesque but unclean wynds, and transforming them gradually, house by house – leaving every inhabitant at home as before, but in a cleaner and more beautiful home, where the air and light come through; grouping friends with friends, and giving them places of reunion for social intercourse and the enjoyment of art. Little by little a whole street, retaining its original character, only without the dirt and the smells, comes out fresh and crisp, like the flower springing clean beneath the foot without a single sod being stirred around the mother plant.[3]

At the same time, Geddes was laying down interdisciplinary educational markers through his editing of *Viri Illustres*, a celebration of distinguished Edinburgh University teachers and students which marked the tercentenary of the university.[4] Fundamental to Geddes's community endeavour was educational and artistic focus. He faced his self-imposed task of improving a neglected area, not by clearance and rebuilding from scratch but by the rediscovery of inherent civic strength and value. Soon after his marriage in 1886 he moved house from Princes Street to live in James Court in the Old Town with his new wife, Anna Morton. Anna was a talented social thinker in her own right who had considered working with Octavia Hill in London in 1882.[5] She became a key point of organisational stability in Geddes's many schemes, and continued in that central role until her death in 1917. Anna and Patrick began renovation and conservation of their own immediate area of tenements as a community-led project with municipal and philanthropic help. The whole notion of conservation and regeneration of urban areas owes much to the initiative they took to rehabilitate those run-down Edinburgh tenements.

Another notable figure in the early years of the Edinburgh Social Union was Elizabeth Haldane, who saw housing work as 'but one side of the bigger work of making life in Scotland, as a whole, more interesting and beautiful'.[6] Haldane had encountered Geddes by chance at Octavia Hill's house in London in 1884, where he was heading a deputation – consisting of himself and his future brother-in-law, James Oliphant – from the nascent Edinburgh Social Union. Although Haldane was only twenty-one at the time, she was persuaded by Hill to learn her system of property administration and to apply it to conditions in Edinburgh. In her autobiography she comments: 'I don't think the immense changes that occurred amongst the more serious part of the community during the eighties have been sufficiently realized ... this was the beginning of all sorts of so called philanthropic movements with a different orientation from the past.'[7] The difference Haldane notes was that movements such as the Edinburgh Social Union, however much they may have owed to the debates within the churches, were no longer dominated by the churches in their organisation. As with so many involved in the Edinburgh Social Union, Haldane had other talents, and in her interdisciplinarity she can be considered another notable example of the Scottish generalist tradition. She was a distinguished translator of Descartes and she wrote the biography of the early phenomenologist James Frederick Ferrier (remembered today for coining the term 'epistemology').[8] Her reflection on life in nineteenth-century Scotland, *Scotland of our Fathers*, is valuable not least for the clarity of its diagrammatic representation of the secessions and unifications of the Presbyterian churches.[9]

Geddes convened the decoration committee of the Edinburgh Social Union and Haldane convened the housing committee. Those two committees were fundamental to the initial structure. Haldane's dealt with acquiring, financing and maintaining accommodation, Geddes's with making aesthetic interventions where appropriate, both within Social Union properties and by facilitating commissions in other locations such as hospitals and church halls. For example, in the first annual report of the Edinburgh Social Union reference is made not only to Phoebe Traquair's work in progress for the mortuary chapel of the Sick Children's Hospital, but the completion of murals based on fairy story images by Walter Crane in the Courant Children's Shelter, and life-size replicas of John Everett Millais's illustrations of the Parables for the Robertson Memorial Church in the Grassmarket.[10] Millais's illustrations would have been familiar to many of Geddes's generation in Scotland, for they were originally commissioned to accompany the retelling of the parables by Thomas Guthrie and published in Norman Macleod's *Good Words* in the 1860s. Guthrie was a key figure in the Free Church of Scotland, and a founder of the ragged school movement. Other artists whose works were copied in that early phase of the Edinburgh Social Union included the German Nazarene painter Overbeck and the English Pre-Raphaelite painter Burne-Jones; indeed, in the second annual report Burne-Jones was thanked for his friendly counsel and it was noted that some of the panels based on his work had been exhibited independently.[11]

The first membership list of the Edinburgh Social Union makes interesting reading.[12] It brings together a group of citizens many of whom continued to support Geddes's initiatives when they outgrew the possibilities offered by the Social Union. The incoming general committee for 1885, as listed in the first annual report, included H. Bellyse Baildon, Gerard Baldwin Brown, Robert C. Buist, Sydney Mitchell and Alexander Whyte.[13] Baildon's links to Geddes included acting as editor for the republication of Geddes's 1884 essay 'John Ruskin: Economist' in *The Round Table* in 1887.[14] That publication also contained essays on 'Ralph Waldo Emerson: Man and Teacher', 'George Eliot: Moralist and Thinker', 'Walt Whitman: Poet and Democrat', 'Charles Darwin: Naturalist' and 'Dante Gabriel Rossetti: Poet and Painter'. It is not clear who wrote the essay on George Eliot, but the authors of four of the essays (Baildon himself on Emerson, Geddes on Ruskin, J. T. Cunningham on Darwin and P. W. Nicholson on Rossetti) appear in the first membership list of the Edinburgh Social Union.[15] The author of the Whitman essay, J. M. Robertson, was not a member; he left for London soon after graduating from Edinburgh University to become assistant to Charles Bradlaugh, editor of *The National Reformer*.[16] In due course Robertson became a distinguished politician. *The Round Table* thus gives an early insight into the intellectual generalism and pedagogical intent typical of Geddes's milieu. It was published in the year of the establishment of Geddes's University Hall and one can imagine the student residents reading such interdisciplinary material with enthusiasm. Baildon was also to contribute to Geddes's magazine *The Evergreen* in 1896.

Of the Social Union members noted above, Gerard Baldwin Brown, first holder of the Watson Gordon Chair of Fine Art at the University of Edinburgh, was another significant Geddes supporter, as was Geddes's Dundee friend, Robert C. Buist. At the time of the foundation of the Edinburgh Social Union, Geddes had recently failed to gain the chair of biology at University College Dundee, which went instead to D'Arcy Wentworth Thompson (see chapter four), but Geddes would in 1888 himself be appointed professor of botany at that institution. In the first annual report of the Edinburgh Social Union, Buist gives his address as 'Well Court', a building which introduced the values of the Arts and Crafts movement to the heart of Edinburgh's Dean Village. The next obvious Geddes supporter on the general committee was the architect of Well Court, Sydney Mitchell, who would a few years later be one of the architects of Geddes's Ramsay Garden complex. Another of those listed, the Reverend Alexander Whyte, is also of interest here. He was a highly regarded preacher, a minister and in due course moderator of the Free Church of Scotland. An outstanding educator (he became principal of New College, now the faculty of divinity of the University of Edinburgh), he would publish widely on subjects ranging from Bunyan's *Pilgrim's Progress* to the mysticism of St Teresa of Avila, John Law and Jacob Boehm. His generalism extended to ecumenicism; indeed, he had a cordial and on occasion moving correspondence with Cardinal Newman and published a selection of his work.[17] With respect to the arts, Whyte had been an advocate of the work of the young Robert Louis Stevenson as early as 1880,[18] and he was a close friend of the outstanding visual artist of the Edinburgh Social Union, Phoebe Anna Traquair, who produced a set of etchings illustrating his lectures on Dante.[19] Almost thirty years later, both Whyte and Geddes would play central roles in welcoming the leader of the Baha'i faith to Edinburgh (see chapter thirteen).

Geddes's initial role as convener of the decoration committee underlines his commitment to visual art. He emphasised that in two remarkable booklets both entitled *Every Man His Own Art Critic*, which were responses to the international exhibitions in Manchester in 1887 and Glasgow in 1888.[20] Towards the close of the second essay, Geddes writes that 'possibilities of new action are thus nearing us; every man may be more than merely his own art critic, but something of his own artist also, for an ideal is returning to animate the labour of his own weary brain and hand'.[21] That statement is a clear reference to the Arts and Crafts Movement, which would in 1889 hold its second congress in Edinburgh. Geddes's words echo John Ruskin and prefigure Joseph Beuys, while at the same time advocating Geddes's own ideal of a society of informed individuals acting in co-operation.[22] Peter Hall comments that it was from 'Reclus and Kropotkin, and beyond them from Proudhon' that Geddes 'took his position that society had to be reconstructed not by sweeping governmental measures like the abolition of private property, but through the efforts of millions of individuals'.[23] That cooperative, anarchist vision was international in reference but Scottish in its moral origin. John P. Reilly has described Geddes as 'a moralist, deeply concerned with bettering man and his lot, before he was a biologist, an anarchist – or a city planner'.[24] Geddes's Free Church background was a crucial factor and it is only if that is recognised that one can make sense of the enduring links Geddes established during the early years of the Edinburgh Social Union with Free Church figures such as John Kelman,[25] Alexander Whyte and Whyte's wife, Jane Barbour.[26] Reilly has illuminated the value that Geddes placed on the Free Church further in his discussion of letters between Geddes and the Russian religious activist, then resident in Edinburgh, Prince Nicolas Galitzin. In one of those letters, in which Geddes is critiquing Galitzin's negative view of anarchism, he underlines the Scottish ideological roots of his sympathy for anarchism by making the intriguing claim that Thomas Chalmers, the founder of the Free Church, was an anarchist economist 'beside whom Kropotkin and Reclus are mere amateurs'.[27] In the same letter Geddes also makes clear his own commitment to the Free Church, asserting that it is the organisation that he is 'proudest of all' to belong to.[28]

But whatever it may have owed to thinkers who were members of the Free Church, the Edinburgh Social Union was in no sense a Free Church organisation. For example, Phoebe Traquair was an Irish Episcopalian by background and Geddes's advocacy of visual art as fundamental to the work of the Social Union found its most profound expression through her murals. She had been born Phoebe Anna Moss in Dublin in 1852. There, in 1873, she married the Scottish palaeontologist Ramsay Traquair, who was that same year appointed Keeper of Natural History at the Museum of Science and Art in Edinburgh.[29] In Edinburgh her art was given focus by the twin influences of Geddes and John Miller Gray. The latter, who became one of her close friends, was appointed first curator of the Scottish National Portrait Gallery in 1883. Writing in the *Art Journal* in 1900, James Caw noted that 'through the stimulating influence of the one and the practical friendship of the other, she came to find the mediums best suited for the expression of her very individual gifts'.[30] She was an Arts and Crafts practitioner in a complete sense and some indication of the admiration she enjoyed comes through her friend W. B. Yeats's comment that 'I have come from her work overwhelmed, astonished, as I used to come from Blake, and from him alone'.[31] Traquair received one of the first commissions from the decoration committee of the Edinburgh Social Union, namely the mural scheme for the mortuary chapel of the Sick Children's Hospital in Edinburgh. Owing to the responsibilities of motherhood it was her first professional commission.[32] The murals were carried out in 1885 and 1886, and subsequently, when the hospital moved to Sciennes, a new version was painted from 1896 to 1898. It included a significant amount of work transferred from the original site, and the fact that such a difficult process of transfer was even attempted gives some indication of the high regard in which the paintings were held. The entire process can be followed in the annual reports of the Edinburgh Social Union. The pride felt by the committee in the initial scheme is clear from the second annual report: 'it is not too much to say that this beautiful and original work is a valuable permanent possession to the city of Edinburgh, and it is much to be desired that it should be more widely known'.[33] Further indication of the high regard in which these works were held is to be found in the *Scottish Art Review* of 1889. There Gerard Baldwin Brown published an article about the murals sponsored by the Edinburgh Social Union in which he devoted a large part of his piece to Traquair's work, and described her as 'sympathetic and thoughtful'.[34] Earlier in the same piece he had noted that the incentive came from Geddes, 'who set himself the task of securing ... interesting and instructive pictorial decoration for the interiors of certain halls and mission rooms in the crowded parts of the town'.[35]

Stylistically Traquair's murals have echoes of Rossetti and Burne-Jones in particular and one of her mortuary chapel panels includes portraits of both.[36] Those portraits are an interesting element of the overall scheme. They are found in small ovals that form part of a series of images of those from whom she drew inspiration. They form the border of her image of *Three Divine Powers* on the south wall of the chapel.[37] Included are Thomas Carlyle, John Ruskin, Alfred Tennyson, Robert Browning, William Blake, Edward Burne-Jones, George Frederic Watts, Dante Gabriel Rossetti, David Scott, Joseph Noel Paton, Dr John Brown, John Miller Gray, Norman Macleod, William Sharp and William Holman Hunt. Early in the history of that mural scheme the subjects of those portraits were no doubt well known but since the publication of a guide, probably around 1900, there has been some confusion that has been repeated in later works. The guide only identifies eleven of the fifteen figures (from Carlyle to Brown) and identifies one of them wrongly. The mis-identification is David Scott, who is identified as his brother William Bell Scott.[38] As a Scottish inspirer of Pre-Raphaelitism and a contributor to Rossetti's *The Germ*, William Bell Scott would have been a most appropriate inclusion. However, David Scott is even more so: Traquair places his image beside that of Rossetti, and above those of Watts and Blake. That echoes Rossetti's view of David Scott as a true successor to Blake, as put forward in his supplementary chapter to Alexander Gilchrist's life of Blake. It is at the same time a recognition of Watts's debt to Blake. Those not mentioned in the guide deserve attention. The inclusion of William Holman Hunt, the most overtly Christian of all the Pre-Raphaelites, makes complete sense in the context of a chapel. The other three figures are less well known today. William Sharp was a talented critic and essayist from Paisley who had made his reputation in London but maintained strong Scottish links. His friendships included Rossetti; indeed, the older man was something of a mentor to him. Around the time Traquair was adding Sharp's portrait to her mural, he was editing *Sonnets of this Century* (1886), which not only included examples by Rossetti but was dedicated to his memory.[39] A further link to Traquair can be perceived here. The publisher of those sonnets was the Newcastle- and London-based firm, Walter Scott, which the next year published a remarkable collection of poetry, *Women's Voices*, edited by William Sharp's wife, Elizabeth. It had the subtitle 'an anthology of the most characteristic poetry by English, Scotch, and Irish Women', and it was Phoebe Traquair who provided decorations for both the cover and the title page. In the next decade both William and Elizabeth Sharp were to play important roles in the literary aspect of the Celtic revival in Scotland, roles which were to be facilitated by Patrick Geddes (see chapter seven). The next figure portrayed in Traquair's mural is Dr Norman Macleod, one of the key Presbyterian activists of an earlier generation (he had died in 1872). His magazine *Good Words* had asserted the integrated role of art and literature in a Christian context from 1860 onwards, not least in its publication of Millais's illustrations of the parables, used, as I have noted, as models for murals by the decoration committee of the Edinburgh Social Union. The final figure in this neglected group is Phoebe Traquair's close friend, John Miller Gray. Mention of Gray reminds one that his appointment as first curator of the Scottish National Portrait Gallery had been supported by two referees who also have a relevance to Traquair. One was Robert Browning, the other was Walter Pater. Not only was Browning the subject of one of the portraits in Traquair's chapel mural, but the poetry of his wife, Elizabeth Barrett Browning, was the basis for one of Traquair's finest illuminated manuscripts, *Sonnets from the Portuguese*, dating from 1895. Pater's influence on Traquair was equally significant and is manifested most clearly in a set of four embroidered panels begun in the 1890s, *The Progress of a Soul*, based loosely on Pater's story *Denys l'Auxerrois*.[40] Note also Traquair's portrait of Dr John Brown (1810–82), author of *Rab and his Friends* and *Horae Subsecivae*. The Scottish writer was one of Ruskin's friends from the 1840s onwards. Indeed, in *Praeterita* Ruskin calls him 'the best and truest friend of all my life; *best* for me, because he was of my father's race and native town; *truest*, because he knew always how to help us both'.[41]

In due course Traquair went on to decorate the song school of St Mary's Episcopalian Cathedral in Edinburgh. Although not a Social Union commission, it was made possible by a Social Union member, the sub-dean of St Mary's, Dr Cazenove.[42] Walter Crane met her around that time and called her work 'delightful in colour and invention'.[43]

Later Traquair made an illuminated book based on the medallions from this scheme. It was bound by the Arts and Crafts bookbinder T. J. Cobden-Sanderson and was made as a gift for Alexander Whyte's niece, Charlotte Barbour.[44] In 1892 the Social Union commissioned Traquair to decorate Robert Rowand Anderson's Catholic Apostolic Church in Mansfield place.[45] The extraordinary nature of her achievement there is reflected in the tenth annual report, in which it is noted that 'the arch leading into the chancel is about 60 feet high by 28 feet wide' and reports 'with very great satisfaction' the completion of the decoration of that chancel arch. Later the hope is expressed that other architects will follow the example of Rowand Anderson and design buildings 'in which mural paintings may be seen to advantage' so that 'art may once more become, as it was of old, the heritage of the people, and not merely the possession of the few'.[46] That report was written in 1894 and as early as 1899 Traquair's work on the church had been noted and illustrated for a popular readership in Katherine Lockie's book, *Picturesque Edinburgh*.[47]

All this gives an indication of the high level of work made possible by the decoration committee and the initial impetus given to it by Geddes. The early structure of the Edinburgh Social Union had evolved by 1889: the decoration committee had merged with another, the art classes committee. The new committee was co-convened by Gerard Baldwin Brown and D. J. Vallance. Although beginning to devote his energies elsewhere, Geddes took his place on the committee along with a number of interesting new members. They included Francis Newbery, who had been appointed director of Glasgow School of Art in 1885.[48] Perhaps it was at that time that Geddes first heard the name of Newbery's promising student, Charles Rennie Mackintosh, from whom Geddes was later to commission designs (see chapter thirteen). Another addition to the committee from Glasgow was James Mavor, professor of economics at Glasgow University and editor of *The Scottish Art Review*. Several painters also became members of the merged committee at this time. One was William McTaggart, who was in the process of radicalising the practice of painting in Scotland. Others involved included James Lawton Wingate, George Reid, soon to be elected President of the Royal Scottish Academy, William Hole, who would go on to make murals for the Scottish National Portrait Gallery, and William Darling McKay, who was both a painter and a pioneering historian of Scottish art.[49] Also on the committee were the architects Sydney Mitchell, Stewart Henbest Capper and David MacGibbon. Mitchell and Capper went on to collaborate on the design for Geddes's Arts and Crafts complex of Ramsay Garden. MacGibbon, along with his partner Thomas Ross, had just published the major three-volume account of *The Ecclesiastical Architecture of Scotland*, and the five-volume *Castellated and Domestic Architecture of Scotland* was in process.[50] Another member of the art classes and decoration committee was David Watson Stevenson, a key figure in the development of public sculpture in Scotland in the wake of the opening-up of that area by Sir John Steell. Steell is most remembered today for his image of Sir Walter Scott, the central element of the Scott Monument in Edinburgh. Some fifty years later, Stevenson had an equally culturally prominent commission, the statue of William Wallace for the exterior of the National Wallace Monument near Stirling. Stevenson had significance for the Social Union both as a supervisor of sculpture classes and as landlord for its studio space. His activity as a sculptor illuminates the cultural nationalism that informed Geddes's thinking. Stevenson's statue for the Wallace Monument was erected in 1887. In the next decade he was closely involved in marking the centenary of the death of that archetypal cultural nationalist, Robert Burns, notably with his statue of the poet's lover Highland Mary, erected in 1896 at Dunoon, and his statue of Burns himself erected in Leith in 1898. A key project of the late 1890s was the making of figures for the exterior of the Scottish National Portrait Gallery, a building funded by yet another Edinburgh Social Union member, John Ritchie Findlay, proprietor of *The Scotsman* newspaper.[51] Stevenson was one of a remarkable group of artists working on this project. His statues for the façade included the geologist James Hutton and Napier of Merchiston, the inventor of logarithms.[52] The images of William Wallace and Robert Bruce were sculpted by William Birnie Rhind. Internally, mural decoration was also important, and the commission was offered initially by the Social Union to Phoebe Traquair, but she turned it down.[53] That decision is consistent with her clear preference for spiritual rather than historical subject matter. The mural scheme, including a frieze of over one hundred and fifty famous Scots, was eventually carried out by William Hole, who was commissioned in 1897.[54] Hole was a distinguished artist whose work successfully set the tone for the interior of that cultural nationalist building. He was also a notable book illustrator, not least of the works of Robert Louis Stevenson. His illustrations include *Kidnapped*,[55] a novel which is itself a statement of cultural nationalist pluralism, telling the tale of two ideologically opposed Scots united by circumstance, friendship and, crucially, by the geography of Scotland itself.

JAMES MAVOR AND THE *SCOTTISH ART REVIEW*

The presence of Glasgow-based James Mavor and Francis Newbery in the Edinburgh Social Union in 1889 sheds further light on the wider networks of which members of the Social Union were part. Mavor was editor of the *Scottish Art Review*, to which both Newbery and Geddes were contributors, and all three had close connections with William Morris and other Arts and Crafts activists. From the art and design side, Newbery 'actively encouraged Morris to visit Scotland'.[56] Mavor's links to Morris were also political; indeed, he had been one of the signatories of Morris's Socialist League Manifesto published in the first issue of *Commonweal*, dated February 1885.[57] Another Scottish-based signatory to this manifesto was the Viennese furniture designer and political activist Andreas Scheu, co-founder of the Scottish Land and Labour League, whom Morris had visited in Edinburgh in 1884.[58] Patrick Geddes was himself a member of this radical group for a short time in the early 1880s.[59] In 1886 Geddes joined Morris in a course of lectures delivered in Edinburgh, Glasgow and Dundee; the papers were collected by Geddes's brother-in-law James Oliphant and published under the title *The Claims of Labour*.[60] As Mavor notes, in due course there was some tension between Geddes and Morris;[61] however, Geddes's admiration for Morris is clear in his 1886 lecture, and Mavor also notes the essential continuity of their goals.[62]

The magazine which Mavor edited, *The Scottish Art Review*, had a short but interesting existence. Its origins were in the desire of the painters of the Glasgow School to create a journal that would represent their concerns in a Scottish and international context. Its first editorial was distinguished by the stated intention that practitioners should write about their own arts.[63] But – like many another cultural magazine project – before it was even published it had run into problems both editorial and financial. It began publication as a monthly in June 1888, and from April 1889 James Mavor appears as editor.[64] Mavor's appointment was controversial. Indeed, the painter Robert Macaulay Stevenson saw his editorship as undermining the original purpose of the magazine.[65] Mavor certainly adopted what was, to use his own word, a more 'eclectic' policy.[66] That included, for example, articles by Kropotkin and Havelock Ellis.[67] The Arts and Crafts strand was reflected not only in previews and reviews of the first and second exhibitions and congresses of the Arts and Crafts movement, but in contributions from Walter Crane and Gleeson White, as well as a cover design by Crane and title page decoration by Selwyn Image. Scottish contributions both verbal and visual came from numerous members of the Glasgow School, including James Paterson, Alexander Roche, James Guthrie, Robert Macaulay Stevenson, John Lavery and Pittendrigh Macgillivray, and also from William McTaggart, James Cadenhead, William Hole, W. G. Burn Murdoch, Phoebe Traquair and William Strang. In his autobiography, Mavor goes out of his way to mention the high production values of the magazine.[68] These were in the hands of yet another Edinburgh Social Union member, W. B. Blaikie of T. and A. Constable.

Blaikie had a crucial role in the cultural activity of the time and in due course he took on the production of works associated with both Patrick Geddes and Phoebe Traquair. For example, in 1890 T. and A. Constable designed and printed the book by Traquair that accompanied Alexander Whyte's lectures on Dante. Elizabeth Cumming has drawn attention to the particular importance of that book in the development of Blaikie's design ethos.[69] Blaikie's main printing work for Geddes was *The Evergreen*, of which more below, but he was in addition a strong supporter of Geddes's projects in the Old Town of Edinburgh. Blaikie was responsible for a number of other exceptional publications involving members of Geddes's milieu. These included the centenary edition of the works of Robert Burns published in 1896 and 1897, and the Edinburgh edition of the works of Stevenson, published as something of a memorial to Stevenson between 1894 and 1898.[70] Blaikie was Stevenson's cousin, and each volume was initialled by him. The first had as its frontispiece a portrait of Stevenson by William Hole, who was also the key illustrator for the *Centenary Burns*. In 1890 a volume which reflected Blaikie's extensive antiquarian interests, *Scottish National Memorials*, had been published, while in 1900 the finest of all books of the Scottish Celtic revival, Alexander Carmichael's *Carmina Gadelica*, was produced under Blaikie's supervision. In 1909 Blaikie became president of the Old Edinburgh Club. The early volumes of *The Book of the Old Edinburgh Club*, a key source for anyone concerned with the history, conservation and regeneration of the Old Town of Edinburgh, are further examples of his work at T. and A. Constable.[71]

Much can be gleaned from the pages of *The Scottish Art Review*. For example, the October 1889 issue includes a review of Thoreau's *A Week on the Concord and Merrimac Rivers*.[72] The anonymous reviewer notes 'the tendency which seems to be developing itself towards simplification of life' that 'stands in definite relation to the influence of Thoreau'. The ecological thrust of that review brings Geddes to mind. It also resonates with the well-established interplay of botany and art during this period, a relationship fundamental to Arts and Crafts thinking. One can note an early statement of that in 1860 in part one of volume five of Ruskin's *Modern Painters*, 'Of Leaf Beauty'. Most of those in Geddes's milieu either operated in both art and science or had very close associates on the other side of the disciplinary line. Phoebe Traquair's husband was an outstanding palaeontologist. Geddes himself, as we know, was a botanist, an ecologist and geographer as well as an advocate of the visual arts. His close associate and former student J. Arthur Thomson was a noted biologist but also wrote key texts to accompany John Duncan's murals at Ramsay Lodge (see chapter six). Similarly, James Cossar Ewart, professor of biology at the University of Edinburgh, was at the same time convener of the Edinburgh Social Union handicraft school committee.

THE ARTS AND CRAFTS CONGRESS OF 1889

The issue of *Scottish Art Review* which includes the review of Thoreau also notes the approaching Arts and Crafts Congress to be held in Edinburgh at the end of October 1889, and the wider context of this event, the second Arts and Crafts exhibition, running from early October to early December.[73] In January 1889 the *Scottish Art Review* had noted the overall success of the Liverpool Arts and Crafts Congress the year before. Geddes's presentation at that first congress, 'Economic Arguments for the Encouragement of the Fine Arts', attracted the interest of the reviewer as one of a group of papers 'valuable for their luminous suggestiveness'.[74] Geddes was also a speaker at the Edinburgh Congress. Some wider sense of Edinburgh during the congress is given by Walter Crane when he writes in his autobiography that:

> among the interesting and sympathetic people one met at Edinburgh must be mentioned Professor Patrick Geddes, whom I visited in his tower and who showed me his schemes for saving the fine old palaces, turning them into residential flats; and we exchanged ideas about the beautification of modern towns, schools, gardens, flowers, and other things.[75]

At the congress, Crane spoke on book decoration and illustration. At the time the Kelmscott Press was not yet in being, but Crane recalls how the socialist group (consisting of himself, Morris, Cobden-Sanderson and Emery Walker) went over from Edinburgh to lecture in Glasgow, and while staying there they discussed the possibility of establishing the press.[76]

The meeting place for the Edinburgh Congress was Robert Rowand Anderson's as yet unfinished Scottish National Portrait Gallery and

Museum of Antiquities in Queen Street. Along with key Arts and Crafts figures from south of the border, including Morris, Crane, J. D. Sedding, Emery Walker, Cobden-Sanderson and C. R. Ashbee, the congress brought together as speakers many of those involved in the Edinburgh Social Union and in the *Scottish Art Review*. The congress has been described as 'one of the most important, and certainly the most illustrious, gatherings in the entire history of the British Arts and Crafts movement'.[77] That comment properly characterises a programme which was a remarkable reflection of the intellectual and aesthetic extent of the Arts and Crafts movement, and of its strong Scottish dimension.[78] The painting section included William Hole speaking on 'Art and the People', W. D. McKay on 'Traditional and Modern Methods of Oil Painting' and J. Lawton Wingate on 'Apprenticeship in Picture-Making'. D. W. Stevenson spoke in the sculpture section, Rowand Anderson was president of the architecture section and David MacGibbon spoke on 'The Characteristics of Scottish National Architecture' while G. S. Aitken spoke on 'The Architectural Education of the Public'. In the applied art section, along with contributions from Morris and Crane were papers by Thomas Bonnar and W. S. Black. Gerard Baldwin Brown was president of the museums section. The full title of that section gives a better idea of its purpose: 'Museums and National and Municipal Encouragement of Art'. Both Joseph Anderson (director of the Museum of Antiquities) and Patrick Geddes were speakers. Typically international in reference, Geddes's topic was 'National and Municipal Encouragement of Art upon the Continent'. A number of combined meetings of sections also took place, and further contributors included the painter Alexander Roche and the architect James Gowans. In the combined meeting of the museums section and the applied art section, Mary Burton spoke on 'Art Teaching in Elementary Schools' while Francis Newbery spoke on 'The Place of Art Schools in the Economy of Applied Art'. Newbery was followed by C. R. Ashbee speaking on 'Decorative Art from a Workshop Point of View'. Ashbee would remain linked with Geddes throughout his career, not least through their mutual friend Ananda Coomaraswamy, of whom more in chapters twelve and thirteen. In the combined meeting of the sculpture and architecture sections, papers were given both by the London architect J. D. Sedding, whose words would soon inspire the young Charles Rennie Mackintosh, and by Mackintosh's then employer, the Glasgow architect John Honeyman.

That period from the foundation of the Edinburgh Social Union in 1884 until the Edinburgh Arts and Crafts Congress of 1889 is the first period of Geddes's interdisciplinary maturity. By the end of it the key network for the development of his career was in place. But part of that network had its heart not in Edinburgh but in Dundee, and not in arts and crafts but in botany.

4

Patrick Geddes, D'Arcy Thompson and the Early Years of University College Dundee

THE DUNDEE NATURALISTS' SOCIETY

IN CHAPTER TWO I noted the significance of Dundee to Patrick Geddes's early higher education. As his projects began to develop he maintained his links to the city, in particular through the Dundee Naturalists' Society. That society played a key role in the dissemination of scientific knowledge in the city. Its commitment to the advocacy of cutting-edge research is clear from a lecture given by the twenty-seven-year-old Geddes in December 1881. The subject was 'The Classification of Statistics'. In the 1880s the nature and use of statistics and their wider implications for the structure of the social sciences were matters of intense debate, not least with respect to their potential for providing a scientific foundation for the nascent discipline of sociology. Geddes's talk was based on a presentation he had made earlier in the year to the Royal Society of Edinburgh. In his own words, 'it was probably the first [paper] which has attempted to organise the whole body of our recorded social knowledge into a form presentable to the cultivators of the preliminary sciences'.[1] The ideas that Geddes discussed at that meeting impressed Geddes's friend J. Martin White.[2] In due course White, who had inherited industrial wealth, was to become a key source of finance not only for Geddes in Dundee and Edinburgh but for the development of sociology as an independent discipline in London. Some thirty years after that Dundee meeting, White financed both of the first two chairs of sociology at the University of London. It may seem surprising that a meeting held in Dundee before it even had a university should have had such significant academic consequences but it certainly gives insight into why Dundee was so much in need of its own university at that time. Patrick Geddes was thus firmly part of the interdisciplinary academic network that enabled the foundation of University College Dundee.

It is likely that Geddes's invitation to speak at the Dundee Naturalists' Society was initiated either by Martin White himself or by the honorary secretary of the society, Frank Young.[3] Young may also have attended Geddes's lectures at the Royal Society of Edinburgh. Geddes had been elected Fellow in 1880 and Young himself was elected in 1882. His proposers included Geddes and the mathematician George Chrystal. Chrystal was another Scottish generalist thinker of note. He was professor of mathematics at the University of Edinburgh, and a biographer and editor of William Robertson Smith. Geddes again lectured to the Dundee Naturalists' Society in February 1882 on the subject of the occurrence of chlorophyll in animals, a topic on which he had made a pioneering research contribution.[4] That biological topic is at first sight in marked contrast to the sociological subject of Geddes's first paper, yet it shares important features with it, for both papers were pioneering and both had a strong interdisciplinary slant. As such they were typical of Geddes, and point to his ability to think across disciplinary boundaries. At the very end of *The Democratic Intellect*, George Davie writes of Geddes as 'forward looking', an apposite description.[5] Geddes was elected an honorary member of the Dundee Naturalists' Society later that same year.[6]

Dundee's industrial growth in textile manufacturing and shipbuilding had been driven by engineering innovation and financial expertise but the strength of that pragmatic culture had not been properly echoed in the availability of higher education. In December 1881 the situation changed completely with the signing of the Deed of Endowment of University College Dundee by Mary Ann Baxter and John Boyd Baxter. That act of educational will is one of the most important single moments in the city's history. In 1883 the new University College enrolled its first students. Towards the end of 1884, applications were invited for the chair of biology. Given his close links to the city, it is not surprising that Geddes, strongly supported by his Dundee friends, applied for the job. In the light of some of the letters sent to him at the time from members of the Dundee Naturalists' Society and others, it seems reasonable to regard him as the front runner. He had the strong support of Frank Young; he had the ear of the College Principal, William Peterson.

However, in the event it was not Patrick Geddes but D'Arcy Wentworth Thompson who was appointed. This was an inspired choice on the part of University College, reclaiming a native of Edinburgh from Cambridge, to give Dundee (and more than thirty years later, St Andrews) the benefit of the teaching and research of one of the great zoologists of the twentieth century. But whatever the benefits of Thompson's appointment, the issue which must be addressed here is why Geddes, with his influential local support, failed to gain the chair for which he was such an obvious candidate. A factor may have been that Geddes, although he had studied with Huxley in London, and had even come into direct contact with Darwin, had never taken a formal degree. That could have counted against him at a time when the sort of codification of achievement in qualifications which we take for granted today was becoming more important. But against such an interpretation is that Geddes was already a Fellow of the Royal Society of Edinburgh, and that Geddes's friends in the Dundee Naturalists' Society, while not regarding his appointment as a foregone conclusion, were at first both enthusiastic in their support and confident of the outcome. An alternative possibility has been noted by Paddy Kitchen.[7] The suggestion is that Geddes's failure to be appointed was at least in part due to doubts raised about his religious orthodoxy. This bears further examination.

On 30 November 1884 Frank Young found himself having to write to Geddes about rumours that were circulating about his religious tendencies, specifically his 'seeking to eliminate the spiritual element from the natural world'.[8] It seems then that, at least in some quarters, his training with Huxley may have counted against him. This is more intriguing when one sees it in the context of the Deed of Endowment of University College Dundee. A 'fundamental condition' of that deed is 'that no student, professor, teacher or other officer . . . shall be required to make any declaration as to his or her religious opinions or submit to any test of [them]'.[9] The irony of finding his own hopes of appointment undermined by what were supposed to be his religious opinions was not lost on Geddes. He responded with bitter humour to a letter from one of his Dundee supporters, James Cunningham, who had requested 'a short statement of your philosophical and theological standpoint, that I might use or not at discretion'.[10]

The underlying question that must be asked is why were Geddes's religious views an issue with respect to this appointment? It may be naïve to think that they would have no bearing at all in any circumstances, even in the light of the Deed of Endowment. But Geddes himself was not making an issue out of them, so why should anyone else have been? The answer to this conundrum, which saw a major appointment to a newly established institution of higher education made in a context which on the face of it seemed to be at variance to the Deed of Endowment of that institution, lies not in Dundee, but in Aberdeen, and not in biology, but in the study of the Bible.

As I noted in chapter one, in the late 1870s and early 1880s the religious and academic structures of the Free Church of Scotland had been shaken and split by the writings of William Robertson Smith (1846–94). In 1881 Smith had been removed from his post as professor of Hebrew and Old Testament Exegesis at the Free Church College of Aberdeen, on grounds of heterodoxy. The issue in question was the 'higher criticism', described, in the words of Robertson Smith's own inaugural lecture, as 'the fair and honest looking at the Bible as a historical record, and the effort everywhere to reach the real meaning and historical setting, not of individual passages of the Scripture, but of the Scripture records as a whole'. He goes on to say, 'This process can be dangerous to faith only when it is begun without faith – when we forget that the Bible history is no profane history, but the story of God's saving self-manifestation.'[11] G. F. Barbour has pointed out that:

> for over five years after his appointment in 1870 Smith quietly pursued his scholar's way in Aberdeen. General statements as to the need for unfettered historical study of the books of the Bible themselves, rather than of traditions regarding them, did not unduly alarm the orthodox, although the more thoughtful might have seen that the postulates of such a critical study were radically opposed to the traditional view of the Bible as of equal historical value in every part. Criticism and a theology based on the idea of literal inspiration could not long exist together.[12]

Those stresses become evident on the publication of Robertson Smith's article, 'Bible', in December 1875 in volume eight of the ninth edition of the *Encyclopaedia Britannica*. To be aware of how this might have affected Geddes one should note Barbour's emphasis on the public and indeed popular dimensions of the debate which took place from 1876 to 1881:

> As the range of the dispute became clear, the whole mind of the Scottish people was stirred to activity and interest. The debates of Presbyteries or Assemblies on the views of Robertson Smith were followed and reproduced in railway carriages and workshops and country smithies.[13]

Far from being an anomaly within the Free Church, Robertson Smith was part of the tradition of investigative interdisciplinary intellectualism that I have noted as being characteristic of it since the church's foundation in 1843.[14] This was no less true in theology than in science or in art. Robertson Smith, although challenging the idea of the literal truth of the Bible, had a great deal of support in that project within the church itself, both in terms of his right to 'follow his scholarly way' and with respect to his conclusions. Indeed, he nearly survived the challenge to his academic role, and he remained a minister of the Free Church for the rest of his life. Nevertheless, what the Robertson Smith case had done was to make issues of the relationship of religion to academia salient just before Geddes's application for the chair of biology was being considered. It might seem unlikely that such considerations would spill over into an appointment in the sciences in a new university established as a secular institution. Yet it is clear from Geddes's correspondence that that is exactly what, at least in his view, had happened. In essence, his right to 'follow his scholarly way' was being challenged.

In his reply to Cunningham, Geddes makes clear his awareness of the direct link between his own predicament and that of Robertson Smith. What at first sight seems to be an obscure aside in which Geddes writes of 'the illustration of the beautiful uniformity of cause and effect offered by the association of Britannica articles with heresy-hunts',[15] is a clear reference to Robertson Smith's removal from his chair in Aberdeen as a direct consequence of his writings for the ninth edition of the *Encyclopaedia Britannica*. Geddes had himself contributed to the same edition and one must conclude from his comment that he believed that, in the light of the Robertson Smith case, this had contributed to undermining his position.

Although Geddes's reference to the Robertson Smith case has not been clear to previous commentators, attention has been drawn to a resonant element indicating his unorthodox religious views, which Geddes's supporters also feared would count against him. This was his friendship with Annie Besant and his disinclination to conceal the fact.[16] Mrs Besant was at that time a major figure in the Secularist Movement, and in that role had been prosecuted in 1877 on grounds of obscenity for the publication of a pamphlet on contraception.[17] She was also a figure of scandal due to her separation in 1873 from her husband, the Church of England cleric Frank Besant. Part of the background to that separation was Annie Besant's ceasing to believe in the divinity of Christ, and consequent refusal to take the sacrament with her husband.[18] Thus Patrick Geddes through his academic and personal contacts both north and south of the border was, at the time of his application for the chair of biology at University College Dundee, associated not only with a heresy case but also with an obscenity trial. In addition, the defendant in the latter had recently declared herself to be at odds with a fundamental aspect of Christian doctrine. This was precisely the sort of free-thinking academic and cultural network in which Geddes thrived, but it can have been of little comfort to him as he saw his hopes of a chair at University College Dundee slipping away.

It is tempting to conclude that had Geddes's position not been brought into question on the issue of religion, he would have been appointed to the chair of biology. Had that chair gone to anyone other than the chosen candidate, D'Arcy Wentworth Thompson, one might feel confident in such a conclusion. But Thompson, though younger even than Geddes, was a biologist of brilliance and as such was well capable of gaining the appointment on merit alone. Geddes himself recognised that merit and despite his own disappointment offered to accommodate Thompson while the latter was transferring from Cambridge to Dundee. In the light of this (and the Edinburgh connections of both men) it seems likely that the two were already friends. In subsequent years they certainly were so. Indeed, in a moving letter to Geddes's daughter Norah, written some years after Geddes's death in 1932, Thompson refers to his lifelong friendship with her father and to his happy memories of Norah herself from her childhood onwards. The occasion of this letter was Norah's desire to have Thompson's comments on a collection of poetry she was proposing to publish. The letter in due course took its place as a kind of informal preface to the book.[19] That evidence of closeness between the Geddes family and Thompson adds a personal dimension to the high regard that Thompson had for Geddes as a thinker.[20]

Whatever the initial disappointment that their candidate had not been appointed, the members of the Dundee Naturalists' Society took to D'Arcy Thompson soon enough. Not long after taking up his appointment in early 1885 he gave his first lecture to the Society on 'Modern Methods of Biological Study'.[21] That year he was also elected to the council. Martin White by this time had taken on the duties of honorary secretary, sharing them with Frank Young.[22] Nevertheless, whatever Thompson's qualities, Frank Young and Martin White must have been concerned by the way in which Geddes had been treated. However, on a positive and unifying note for all concerned, the scientific standing of the Dundee Naturalists' Society itself was underlined that same year of 1885 by the election to fellowship of the British Association for the Advancement of Science of three of its members – Martin White, Patrick Geddes and D'Arcy Wentworth Thompson.[23]

PATRONAGE FOR A PROFESSOR

Geddes's failure to gain the Dundee chair in biology in 1884 seems to have put his career advancement on hold. He had already failed, in 1882, to be appointed to the chair of natural history at Edinburgh University. In 1888 he failed again to be appointed to another Edinburgh chair, for the Regius chair of botany had fallen vacant on the death of his friend Alexander Dickson in 1887. Geddes's problems were compounded by the fact that he was now showing himself to be a strongly interdisciplinary thinker at a time when narrow expertise was beginning to be the index of academic recognition that it still is today. We now take Geddes's generalism for granted; indeed, it is rare to describe him with reference to only one area. But his pioneering and wide-ranging thinking, frequently in areas which were only properly defined as academic disciplines in the wake of his exploratory work, was to confuse his more conventional colleagues for the rest of his career, even though, as I have noted in chapter one, it continued an earlier Scottish generalist tradition. Geddes was capable both of presenting an expert view and of understanding and valuing the wider intellectual and cultural context of such expertise. One of his aims was to develop the academic discipline of sociology as a way of giving structure and method to the study of such wider contexts, issues and sympathies. His papers to the Dundee Naturalists' Society in 1881 and 1882 had already illuminated that breadth, for they reflect both his specialist knowledge of biology and his desire to situate all knowledges within a wider sociological framework. However, that same generalism opened him to accusations of lack of focus and such comments did not help him in his academic career.[24]

In that situation of uncertainty the financial support of Martin White was of all the more importance to Geddes. White had come to control the family wealth on the death of his father in 1884 and by 1886 the relationship of patronage as well as of friendship between Geddes and White had begun to assume a pattern which was to continue for more than thirty years. That consisted on the one hand of Geddes asking White for money for various schemes, and on the other of White attempting to finance Geddes in a way which was both controlled and creative. Not surprisingly, there were tensions between the two men. For example, in a letter of 27 January 1886 White writes from Dundee to Geddes with reference to the terms of a £200 loan requested by Geddes for some unspecified purpose. The tone of the letter is very much that of a close friend; indeed, part of the letter refers to a trip to Greece which the two men were about to make. However, this intended journey also draws attention to the stresses consequent on disparity of wealth, for Geddes's participation in the trip was made possible because he was being employed as a tutor. White continues:

> now regarding going away I assure you I wish you not to feel, and I shall not, the rich and poor sentiment you expressed. If I can put a good scientific

man in 'real good shape' consciousness of accomplished duty is my reward, for I advance science. And you must in this instance remember the pleasure I have in scientific company and in yours especially. You are to be cashier and we are to travel unrestrained and joyously.[25]

The main body of the letter is full of suggestions about Geddes's future, which show White's concern that Geddes's work should not be lost to academia, and the fact that White even mentions the possibility of such loss suggests Geddes's uncertainty of academic direction at this time. It is clear that – quite apart from his role as a patron – White saw himself both as a personal and as an academic advisor to Geddes. For example, without even starting a new paragraph, White shifts from loan repayment details to considering Geddes's prospects:

I think you want some true stimulus to take you out of the in some respects, unproductive speculative and give the world some of your matured thoughts or conclusions. Your publications would be valuable in themselves, and their loss must not be risked.[26]

White goes on to write of introducing Geddes more fully to the public, where he would 'gain an enlarged sphere of usefulness, beneficial influence and enlarged aims towards further work . . . And all this you could do without much sacrifice to your pursuits.'[27] In that passage White seems to be sounding out Geddes with respect to some sort of stable employment. And in due course, after Geddes had failed to be appointed professor of botany at Edinburgh University, White and his siblings financed a part-time chair of botany, specifically for Geddes, and named in memory of their father James Farquhar White, at University College Dundee. It is by no means clear that this scheme was in White's mind when he wrote to Geddes in 1886, but from the tone of the letter it seems likely that White was at least beginning to consider some such idea. One wonders how much an awareness of perceived injustice done to Geddes in the treatment of his application for the chair of biology was a factor in White's suggestion to University College Dundee that such a chair, specifically for Geddes, might be appropriate.[28] Whatever the case, in 1888 Geddes was appointed as professor of botany on an agreed part-time basis which left him free to pursue other interests. Quoting *The College* magazine of December 1888, Matthew Jarron notes that Geddes made an immediate difference to the appearance of University College, transforming gravelled paths and grassy plots into banks of greenery and beds of flowers.[29] Jarron also points out that it was White again, at Geddes's behest, who provided funds for another key employee of the college, the gardener William Watson. He took up his post in

1890 and became a lifelong friend of Geddes. His appointment was crucial for Geddes's teaching activities.

In 1889 Geddes published his first book, *The Evolution of Sex*. Its co-author was his former student in Edinburgh, J. Arthur Thomson, who was to be appointed professor of natural history at Aberdeen University in 1899, a post he held until 1930. Thomson became one of the out-standing writers of biology books – both for students and a general readership – of the early twentieth century. Over forty years after the publication of *The Evolution of Sex*, Geddes's last book, *Life: Outlines of Biology*, published in 1931 was another collaboration with Thomson. Helen Meller has pointed out that *The Evolution of Sex* 'proved to be one of the most influential books for a decade . . . on one of the most burning subjects of the day'.[30] In Geddes's thought it has a special place as an expression of his commitment to evolution as a general theory of explanation, not only as it applied to aspects of biology but to aspects of society. Although a biological text from beginning to end, the last sentence of *The Evolution of Sex* points firmly to Geddes's societal inter-ests. It is a quotation from Schiller: 'While philosophers are disputing about the government of the world, Hunger and Love are performing the task.'[31] It can be seen as preparing the way for Geddes's exploration of urban development, *Cities in Evolution*, published in 1915. It should be stressed, however, that Geddes was not interested in evolution as a biological theory to which sociological data could be reduced. What he was interested in was evolution as a process which could take place in a variety of complex systems, human biology being one of them and human society being another.

Geddes differed from Huxley in that he saw evolution as more a co-operative than a competitive process. For him it was not a question of Tennyson's nature red in tooth and claw. As Stewart Robertson noted:

> I recall the thrill that went through an audience as Geddes traced the basal feature of all life to be the sacrifice of the mother for her offspring, and closed by saying, with the usual fingering of the abundant locks and the phrase over the shoulder, 'So life is not really a gladiators' show [as Huxley had charac-terised it], it is rather – a vast mothers' meeting.'[32]

So by the end of the 1880s University College Dundee found itself with two remarkable biologists – Patrick Geddes and D'Arcy Wentworth Thompson – on its staff. By endowing the chair of botany, Martin White had recognised Geddes's need for a stable position in a university. Geddes had been provided with a firm academic identity which gave coherence to his career at precisely the time it was needed. He began to repay his debt to Dundee immediately; indeed, it is highly likely that it was Geddes's appointment in 1888 that stimulated the professoriate of University College to establish the Dundee Social Union on 24 May of that year, six weeks after he took up his post.[33] But Geddes's greatest contribution to Dundee was in terms of botanical, geographical and ecological research. In the ensuing years (see chapter ten) his Dundee students were to become leaders in the development of both ecology and geography as disciplines.

His growing reputation did not, however, make it any easier to fund his various projects. The correspondence between Geddes and White over the next decades repeats a familiar pattern of loans and friendship, the loans often now directly related to projects related to Geddes's Dundee chair, but also to other projects such as the finishing of his Ramsay Garden project in Edinburgh.[34]

5

Education, Anarchism and Celtic Revival in Edinburgh

UNIVERSITY HALL: BUILDINGS AS INTERDISCIPLINARY EDUCATION

GEDDES'S EXTRAORDINARILY INTERDISCIPLINARY commitment comes into full focus in the 1890s. During that same period in which he was developing his department at University College Dundee as a centre for botanical research, in Edinburgh he moved on from the primarily philanthropic activities of the Social Union to educationally driven projects. At their heart was interdisciplinary thinking, which Geddes facilitated by developing a network of residences and academic facilities which took the overall title of 'University Hall'. In so doing he drew inspiration from the earliest European universities, which had been first and foremost communities of students. For Geddes that integration of academic and student life into the heart of the old city served both his higher educational and civic aims. His home in James Court was close to the site of Ramsay Garden, which he would in due course develop as a complex of private flats and student residences. It was even closer to the Outlook Tower, which would become the intellectual core of his educational endeavour.[1]

This new phase of his work transcended the concerns of the Edinburgh Social Union, shifting the emphasis from remedial social intervention and urban conservation to higher education and the propagation of new world views. Reflecting on this change in emphasis, Elizabeth Cumming comments that Geddes by this time considered the Edinburgh Social Union to be 'out of touch with the urgent issues of social reform'.[2] His new activities led him to reduce his role in the Social Union and indeed caused some irritation among those who had followed his lead into that body in the first place. Boardman is no doubt close to the mark when

he makes the point that it was characteristic of Geddes to initiate a
project and then, having got it up and running, to begin to concentrate
on something else.[3] However, one should note that not only Geddes
but many of his friends remained involved to some degree with the
Edinburgh Social Union for many years, as the membership lists in the
annual reports show. For example, the first person to write a guide to
Geddes's Outlook Tower was John Kelman.[4] He did this in 1905, and
the next year he is recorded in the annual report of the Edinburgh Social
Union as proposing a vote of thanks at the Annual General Meeting.[5]
Geddes himself is listed as a member of the Social Union until 1930, two
years before his death.

   University Hall was founded in 1887. An account published in 1903
records the event, relating how in that year 'one of the younger teachers
connected with the University of Edinburgh', namely Geddes, 'impressed
with the need of providing for the undergraduate and postgraduate students something better and more collegiate than isolated board and lodg-
ings', rented three flats 'where graduates and students might enjoy the
social and other advantages of residence in common. Thus was initiated
the system of Residence now known as University Hall.'[6] Soon the whole
tenement at 2 Mound Place was taken over, and in 1889 Riddle's Court
was purchased, expanding the total capacity to forty. That purchase set
a pattern for Geddes of creating modern uses for historic properties, for
Riddle's Court is among the most historically interesting buildings in
Edinburgh, having among its distinguished former residents the philosopher David Hume. While noting Hume's connection to Riddle's Court,
J. B. Salmond commented that 'among all the ghosts that wander here,
the kindliest surely is that of Patrick Geddes, who did so much to keep
this court alive'.[7] Geddes had Riddle's Court altered and renovated in
the early 1890s. The architect was Stewart Henbest Capper, who added
an external staircase very much in an Arts and Crafts mode. The motto
of University Hall, *Vivendo discimus*, that is to say 'living we learn', was
carved above the archway. It is clear that by 1895 Geddes regarded that
motto as something of a definition of his activities. That year he gave a
substantial series of lectures, of which he wrote: 'The whole Course is in
fact an extension of that general motto which you see carved or printed
over our buildings and publications "vivendo discimus".'[8]

   The next element to be acquired for University Hall was the most
important of all and it is indicative of Geddes's ambition as an educator.
This was the purchase in 1891 of the house on Castlehill which had
belonged to the poet Allan Ramsay. Along with it Geddes acquired the
site of what was to be a new informal complex of flats, student accom-

modation and meeting rooms, in due course to be known as Ramsay Garden. By 1893 the poet's old house had been transformed into Ramsay Lodge, a student residence capable of accommodating some forty students. That was to be the heart, both ideological and physical, of the 'accommodation of graduates, extra-mural teachers, and others more or less connected with the University'.[9] It is hard now to imagine Edinburgh without that centrally located synthesis of Arts and Crafts aesthetic philosophy and tenement living. Ramsay Garden is both traditional in ethos and modernist in implication.[10] Scottish and international in its visual reference, it is an essay in turrets, gables and balconies that draws on a multiplicity of sources, not only through direct quotation but through its dominant physical position within the psychogeography of the city. Immediate reference points include the classical integrity of the seventeenth-century Heriot's Hospital to the south, the coherent diversity of the Royal Mile to the east, the proto-modernist grid of the New Town to the north, and the incremented solidity of the Castle to the west. The main architects involved were Stewart Henbest Capper and Sidney Mitchell.[11] Many features of Ramsay Garden hark back to Geddes's commitment to earlier periods of history both Scottish and European. The project set the seal on Geddes's challenge to the decline of the Old Town of Edinburgh and created the conditions for further cultural revitalisation. In the annual report of the Cockburn Association for 1892–3, Geddes was praised for the commencement of the project; indeed, he joined the council of the Cockburn Association in 1894, continuing his involvement until he left for India in 1914.[12] Ramsay Garden was not just a building, it was an educational community, an informal college, built in and around a garden, with reference to a city, with reference to a region. Geddes's commitment to University Hall as an integrated world of students and teachers is implied in his own postal address: 'Ramsay Garden, University Hall, Edinburgh'.

To finance his projects, Geddes obtained funding from a variety of sources. For example, in the 1890s, in a key area of the Old Town adjoining James Court, he persuaded Lord Rosebery to buy the dilapidated building known as Lady Stair's House and to have it renovated to designs by G. S. Aitken.[13] As a consequence it survives today and is now the Writers' Museum, devoted to Robert Burns, Sir Walter Scott and Robert Louis Stevenson. A photograph of its condition prior to renovation appeared in 1910 in *The Book of the Old Edinburgh Club*, illustrating a piece by Geddes's supporter Thomas Whitson.[14] This shows just how derelict the structure had been, and its survival must have been a matter of real doubt before Geddes's intervention. In addition to facilitating acts of private patronage, he entered into a number of contracts with the city of Edinburgh. As noted in chapter three, the task of understanding those arrangements has been considerably eased by the research of Lou Rosenburg and Jim Johnson.[15] Direct private support was also important: Geddes's wife Anna came from a wealthy family and her money was essential backing for many of his projects, as was the patronage of friends such as J. Martin White in Dundee. Geddes also persuaded interested parties to pay in advance for the construction of Ramsay Garden flats. But the realisation of his ideas usually exceeded his budget and, as the University Hall projects expanded, he ran into financial and administrative difficulties. These were to some degree resolved by the foundation of a body, the Town and Gown Association, to manage Ramsay Garden and University Hall, but signs of these problems come through in his correspondence with the artist John Duncan. For example, in a letter of November 1895, Geddes writes reproachfully to Duncan in response to complaints about lack of funding for the Old Edinburgh School of Art and money owed to Duncan. He then tries to reassure Duncan by emphasising, 'I thoroughly appreciate the work you have done and are doing, and I thoroughly value the assistance of the ladies you have trained.'[16] It seems clear that Geddes was taking Duncan's support somewhat for granted. It is equally clear that Duncan continued to give that support, regardless of financial problems.

For Geddes every building was a potential educational manifesto. He was attracted to the Scottish Reformation habit of engraving educative quotations and mottoes above doorways and on buildings. Beneath the windows of the Geddes family flat in Ramsay Garden is a sundial bordered by quotations in the Greek of Aeschylus and the Scots of Robert Burns. [Figure 3.] It dates from 1892. The Aeschylus has been translated as 'Time heals all things in ageing them',[17] the Burns reads 'Its comin yet for a' that'. With these two statements from two poets, one originating in the folk/classical culture of ancient Athens, the other originating in the folk/neoclassical culture of Scotland, Geddes emphasises the mutual illumination – through similarity and difference – of alphabets, of languages, of poets and of periods of history. That sundial thus – as a sundial – evokes the present moment within the present day; and as a bearer of quotations it situates that moment with respect to both the Scottish past and to a culture foundational to Europe, namely that of ancient Greece. When the meaning of the words is taken into account, Geddes is seen to be making the political, cultural and historicist point that we, in the Scottish poet's words, are 'comin yet' to a better era, despite everything. This phrase comes from Robert Burns's song 'Is there, for Honest Poverty', an anthem in the cause of social justice. Written in 1795, it reflected the contemporary cry of 'liberty, equality, fraternity' in France.[18] Bringing the ancient Greek into play, we find the historicist notion amplified. In a translation of the Aeschylus quotation by John Stuart Blackie, it reads 'Time, that smooths all things', a line from *Eumenides* in which we find Orestes caught up in a process of atonement and redemption. Such redemptive historicism resonates with the anarchist thinking of Kropotkin and Reclus, with which Geddes felt close affinity.[19] Geddes's mottoes are reminders that there is no such thing as historical neutrality in everyday life. The phrase 'Ramsay Garden' becomes itself one of Geddes's mottoes. The name may have existed already, but Geddes saw its potential and realised it. 'Ramsay' focuses attention on the fact that these flats and residences have at their core the house of Allan Ramsay the eighteenth-century poet. 'Garden' emphasises the cultivation of nature as the complement of both poetry and architecture. Thus the name Ramsay Garden implies both culture and ecology, and it is the appropriate interplay between the two that Geddes believed gave quality of life in any urban – or rural – situation. That cultural-ecological philosophy, which Geddes saw as the philosophy which should underlie all planning, spread out from this central point in the city of Edinburgh, and manifested itself in the publication of books and the commissioning of art on the one hand and in the creation of gardens from wasteland and the rehabilitation of slums on the other.

A PARTY WITH KROPOTKIN

The sundial provides a gloss on identities – notational, linguistic, national, political, moral, historical, ecological. It is Geddes's intellectual and cultural signature on the exterior of Ramsay Garden. His friendship with Kropotkin helps us to appreciate the human presences which lie behind the sundial quotations. They had met when the Russian was in Edinburgh in late 1886 staying with John Stuart Blackie. Discussing this period, Kropotkin's biographers note Blackie's role as a classical scholar and translator of Aeschylus; indeed, Blackie contributed the article on Aeschylus to the ninth edition of the *Encyclopaedia Britannica*.[20] In Edinburgh, Kropotkin 'made the acquaintance of a number of men who were later to be his closest scientific friends'.[21] Along with Geddes, who is described as sharing Kropotkin's views on many sociological matters, another was James Mavor.[22] In his autobiography, Mavor gives a valuable account of this network of cultural, political and scientific figures. He recalls meeting Kropotkin one evening at a party at Geddes's flat in

James Court.[23] Among those present was a young Norwegian biologist studying with Geddes at the marine biological station at Granton. Fridtjof Nansen would in due course become one the best known of all polar explorers and his presence underlines the fact that for Geddes a necessary part of cultural revival was geographical exploration in the cause of science. Another of Geddes's students around this time was William Spiers Bruce, who signed up in 1887 for Geddes's summer school at Granton.[24] In the early years of the twentieth century, Bruce was to lead the Scottish National Antarctic Expedition. In Geddes's milieu, such exploration enabled the properly comparative understanding of the local. It had little to do with imperialism and everything to do with an understanding of ecology.

Having introduced us to Nansen, in his next sentence Mavor notes the presence at Geddes's party of Thomas Kirkup, author of the article on 'socialism' for the *Encyclopaedia Britannica*.[25] He continues, 'Kropotkin and I alternately hammered Marx and Kirkup spoke up for him, and we had an excellent discussion.' Mavor then accompanied Kropotkin back to his accommodation in John Stuart Blackie's house, laying the basis of a long-standing friendship.[26] It is in the context of such gatherings that Geddes's pamphlet *Co-operation versus Socialism*, published in 1888, must be seen. Here Geddes acknowledges both Kirkup and Kropotkin, and he writes that 'few people adequately realise how good a case can ... be stated for socialism; still fewer know how an even better case can be stated for anarchism; while fewest of all yet recognise in those apparently strange developments of modern thought much of the oldest, commonest, and most enduring wisdom of the human race'.[27] Geddes and Mavor were close contemporaries and good friends. Kropotkin's host John Stuart Blackie was their senior by about half a century but in several respects he set an example for Geddes in particular. Blackie was Professor of Greek at the University of Edinburgh, but in addition he was a strong proponent of Celtic revival. Like Geddes he mixed a romantic Celticism with a cultural pragmatism, and one crucial outcome of his efforts was the raising of funds for a chair of Celtic at the University of Edinburgh.[28] Furthermore, he was a trenchant critic of the Highland Clearances, the enforced emigration of Highland Gaels in order to create sheep farms or deer parks on their land, summing up the iniquities of Highland land use as follows: 'the economical capacities of the Highlands are not to be understood by a few idle young gentlemen from the metropolis, who travel over the bare brown moors for ten days or a fortnight in the autumn, and then conceit themselves that they have seen the country'.[29] He was also an advocate of the poetry of Robert

Burns and, as noted, a translator of Aeschylus. Geddes's contact with Kropotkin while the latter was staying with Blackie thus provides context for understanding the Ramsay Garden sundial. One might even consider the historicist quotations from Aeschylus and Burns as Geddes's tributes to Blackie and Kropotkin, respectively. Blackie died in 1895, three years after the sundial was erected, and Geddes's regard for him is evident in his seminal essay, *The Scots Renascence*, published in *The Evergreen* that year. This begins with an account of Blackie's funeral as an expression of a pluralistic cultural nationalism. A further tribute came in the naming of 'Blackie House', a newly converted tenement that became part of University Hall in 1897. That building was again given a physical signature, not with a sundial but with an oriel window overlooking the Mound and Princes Street Gardens. On it is carved a portrait of Blackie flanked by a Scottish thistle and a Celtic harp.

J. P. Reilly describes Geddes as 'an Anarchist of the Chair', that is to say 'a professor who sympathized with many of the political and social goals of his friends Kropotkin and . . . Reclus, but who chose to achieve these ends by less direct and less attention-seeking means'.[30] Geddes's anarchist views led him to doubt the value of conventional politics in achieving cultural goals. Insight into his view of the ambiguous association between cultural activity and the political establishment of the day, even that part of it concerned with reform, is to be found in a letter written by him in November 1895. It shows that his commitment to cultural revival and the cause of home rule did not translate into a conventional political position in any easy sense.[31] In a reply to Campbell of Barbreck, he declined to join the Home Rule Association and criticised its members for their failure of cultural engagement. 'I am sorry', he writes, 'to decline your invitation but I believe I can do best service to the cause by working at the realities of the Scots Renascence. I believe all the peers and members can do nothing real or permanent until the literary, the academic, the social, the industry movement of Renascence floats them on.'[32] Geddes thus makes very clear that for him political change follows cultural change, not vice versa. He would not have been surprised that the reconvening of the Scottish Parliament in 1999 happened after two decades of intense activity in the arts. He continues:

> It is just in the interests of furthering Scottish literature and other realities that I feel bound to stay away. As an illustration of my total dissent from contemporary political methods I may say for instance that I similarly won't go to temperance meetings not because I don't want much greater temperance, but because I find in practice that the political teetotalers won't come to a real temperance café when I build one and I must wait until I see members of

the home rule association furthering Scottish literature, work for the Scottish Universities, Scottish Art, Scottish industries . . . before I could join them.[33]

THE NEW EVERGREEN

Just as the sundial can be seen as a visual expression of Geddes's thinking on the exterior of Ramsay Garden, in the interiors Geddes's thinking was made manifest in commissioned mural schemes, both for Ramsay Garden and for other parts of University Hall. In 1897 these murals were given prominence in an article for *The Studio* and in 1898 *The Artist* followed suit. In a Scottish context, *The Studio* is normally thought of as an advocate of the circle of Charles Rennie Mackintosh rather than that of Geddes, but there are a number of interesting pieces related to Geddes. These include one of a series of articles detailing different locations as sketching grounds for artists. 'Edinburgh as a Sketching Ground' was published in 1896. It was written by Margaret Armour and illustrated with work by William Brown Macdougall.[34] Armour is perhaps most remembered today for her translations of Wagner's *Ring of the Nibelung*, which were illustrated by Arthur Rackham. Armour's advice to any artist coming to Edinburgh was to live at University Hall: 'You will be well housed, well fed, well amused . . . you will be in stimulating touch with Professor Geddes, a most enlightened lover of the Fine Arts.'[35] There could be no clearer indication of the practical value of Geddes's schemes for artists. Macdougall played a prominent role in providing illustrations for the first manifestation of the mixture of art, science and literature that was to be fully realised in Geddes's magazine *The Evergreen*. That was *The New Evergreen*, its title referring directly to Allan Ramsay's collection of early Scottish poetry, *The Ever Green*, published by Thomas Ruddiman in 1724.[36] The cover of *The New Evergreen* underlines the relationship by enclosing the dates 1724 and 1894 in the stylised root system of a tree of life. This forty-page magazine was produced as a Christmas celebration by residents of University Hall in 1894.[37] It thus predates the first issue of Geddes's *Evergreen* itself by a few months. The introductory 'Essay Commendatory', written by William Macdonald and Victor Branford, is a sort of manifesto for University Hall. It includes the following summary:

> Be this, then, our reading of the aims and uses of University Hall as we know it today: To be a kindly caravanserai for the student from near and far; to be a progressive crystallisation of diverse social elements; to be a Workshop that may one day be considered a School of Art; to be a School of History which begins with the story of Old Edinburgh, and goes in time to the ends

of the earth; and first and last, to be a receiving focus and a radiating centre of all that is good to know and feel.[38]

Geddes did not contribute any of his own writing to *The New Evergreen*, but that statement of internationalism and interdisciplinarity shows how well his message had been understood. There is a general sense of wandering scholars, given a further global push by the 'caravanserai' hint towards *Omar Khayyam*. But note the specific reference to the beginnings of the Old Edinburgh School of Art on the one hand, and on the other to the Outlook Tower, 'the School of History which begins with the story of Old Edinburgh, and goes on in time to the ends of the earth'. With respect to Allan Ramsay's example, the sentiments are echoed and amplified by Branford, writing jointly with Geddes almost a year later, to introduce the second volume of *The Evergreen*. In that *Book of Autumn* they wrote:

> We who inherit Ramsay's old home and would also follow in his steps as workers and writers, publishers and builders, are seeking to gather such traditions as still linger around us, to set down such thought or song as may be in ourselves – hopeful at least of suggesting better things to those who will follow us here.[39]

The University Hall residents responsible for *The New Evergreen* included a number of figures soon to be identified as among Geddes's most significant academic followers and artistic collaborators, in particular J. Arthur Thomson, Victor Branford and John Duncan. Other contributors – William Brown Macdougall, Margaret Armour, William Macdonald, W. G. Burn Murdoch – have already been mentioned. In addition one finds work by James Cadenhead, W. E. Garrett Fisher, William Smith, J. J. Henderson, Gabriel Setoun, C. H. Laubach, Hamilton More Nisbet and Louis Weierter. The last mentioned later included an introduction by Geddes in his *Story of Edinburgh Castle*. That introduction sheds further light on the early days of University Hall. Geddes writes of:

> groups of students, happily mingled with artists, who in hostel after hostel of Castlehill and Lawnmarket and St. Giles were in those days initiating for Edinburgh the adoption of that free and informal mode of associated life from which the historical colleges of medieval universities have been a further but not always a more vital development.

He continues:

> And I speak here first of this little fellowship of University Hall . . . as having set its stamp upon us . . . since common environment and associated action

ever tend toward unity of interest and harmony of spirit. For our association of 'town and gown' . . . was already foreseeing, in this union of artist and student, the educational future, in which the old schism between knowledge and beauty, the long separation of learning and art, and of these from the common life, shall be abated.[40]

Of that 'little fellowship', Thomson, Branford, Macdonald, Armour, Duncan, Cadenhead, Burn Murdoch, Smith, Setoun and Laubach all went on to contribute to *The Evergreen*. Like *The Evergreen*, *The New Evergreen* was printed by T. and A. Constable, and its high quality attracted a positive review in *The Artist* of April 1895, supported by a full-page illustration: *The Three Princesses* by W. Brown Macdougall.[41] On the previous page of that issue of *The Artist* there is news of a work by D. W. Stevenson whose significance as a member of the Edinburgh Social Union was noted in chapter three. The work in question, his statue of *Highland Mary*, was due to be installed at Dunoon in 1896 to mark the centenary of the death of Robert Burns. That year would also see Stevenson casting in bronze John Duncan's design for *The Witches Well*. This sculpted water fountain was made to mark the place where witches had formerly been burnt on Castlehill, close to Ramsay Garden. Duncan would make a contribution to the centenary of Robert Burns as designer of the cover of a centenary edition of the poet's works which also contained a notable suite of etchings by William Hole. The review of *The New Evergreen* in *The Artist* concludes by reproducing a set of aphorisms by Victor Branford, which concluded with a rousing commitment to the generalist ethos of the newly built heart of University Hall, Ramsay Garden: 'the separation of Arts and Crafts is the root of all evil'.[42] The final image of *The New Evergreen* shows the skyline of Castlehill extending from Ramsay Garden to New College. Prominent is the proposed remodelling and heightening of the Outlook Tower in a manner loosely inspired by the Old Netherbow Port, destroyed over a century earlier but of fond memory, not least as rebuilt by Sydney Mitchell as part of the Edinburgh International Exhibition of 1886; however, that final phase of building never came about.

The penultimate image of *The New Evergreen* is perhaps the most interesting of all. It is Geddes's visual signature on the magazine, a collaboration with John Duncan entitled *Lapis Philosophorum*.[43] [Figure 1.] It was used later as the design of a stained-glass window (now lost) in the Outlook Tower, one of two complementary windows made for the tower for pedagogical purposes. The other window showed Geddes's *Arbor Saeculorum*. [Figure 2.] That design became the signing-off work of the first issue of *The Evergreen*. Both are visual explorations of generalist thinking: where *Lapis Philosophorum* is concerned with the unity of arts and sciences, *Arbor Saeculorum* is concerned with the historical evolution of ideas, spiritual and temporal. The broad distinction here is, on the one hand the laws of nature and the methods needed to discover them – set in the philosopher's stone, so to speak; on the other hand are the ever-changing data of human society as it has evolved presented as a tree of the eras of human life. The initials 'P' and 'G' form a prominent part of the base of each design. *Lapis Philosophorum*, then, is Geddes's symbolic signature on *The New Evergreen*, and the same image is used two years later to conclude the entire series of *The Evergreen*. Although not commented on in *The New Evergreen*, this 'philosopher's stone' is described in a guide to the Outlook Tower published in 1906 as an image of 'an obelisk whereon is outlined in graphic notation a classification of the Arts and the Sciences',[44] which sums it up pretty well. It is first and foremost a classification for educational use. Geddes's concern here is with public communication. His primary point is that what we call arts and sciences are deeply intertwined with one another:

> Beginning at the right-hand side of the upper division, we have the musical staff, suggestive of the general idea of notation, and of symbolising the theory of Aesthetics, which underlines all Fine Art. This is represented on the opposite side by the compass and square of the Architect, Architecture being not only 'frozen Music' but the most synthetic of the Fine Arts, since it combines with itself sculpture and painting.[45]

And he continues:

> Next comes the Butterfly, the Psyche, representing Psychology, the subjective aspect of the Science of Life. Opposite is the Flower, from which the butterfly derives its sustenance, indicates the art of Education, whose function it is to provide fit nourishment for the child soul, and which is (or ought to be) based on a true Psychology. In the uppermost division the Celtic Cross with its sun-circle (at once a cosmic and a human symbol) stands for Ethics; for must not a complete science of Ethics take account of our relation to the cosmos as well as humanity?[46]

The passage quoted here comprises less than a quarter of what is written about the *Lapis* image, yet it brings in a number of ideas central to Geddes's thinking, not least the emphasis on graphic notation and the relationship of painting and sculpture in architecture. Again, his human ecologist's vision of ethics needing to take account of our relation to the cosmos as well as to humanity is a direct precursor of the thinking of twenty-first-century human ecologists.[47] And, of course, our relation to the cosmos is mediated by the rhythms of nature, diurnal – as we have seen in the sundial – and seasonal, as Geddes explored in *The Evergreen*, his 'northern seasonal'. The fact that the image that represents Ethics is the sun-circle of the Celtic cross is germane, although it should be noted that in the reproduced version of the *Lapis* that form of the cross does not appear, although the position of the conventional cross represented on the obelisk does have a large sun directly behind it. Indeed, the obelisk is very much represented as the gnomon of a sundial, with the three versions of the sun in three different noon positions. One can take the highest position to represent the summer solstice, the middle position to stand for the two equinoxes and the lower position to indicate the winter solstice. Thus the study of the seasons via knowledge of the cosmos is also symbolised here.

Two sphinxes guard the arts and sciences united in symbol on the obelisk, and these creatures relate also to John Duncan's other visual contribution to *The New Evergreen*, an image accompanied by a poem entitled *Venus Consolatrix*. Despite the poem's title, the image shows not Venus but rather a youthful Oedipus-like figure giving himself up to the embrace of a predatory Sphinx in the shadow not of the Greek city of Thebes but of the Pyramids, thus sharing its Egyptian reference with *Lapis Philosophorum*. Duncan's image may have been influenced by the two figures in Charles Ricketts's *Crouching by the Marge*, made as an illustration for Oscar Wilde's *The Sphinx*, published earlier the same year.[48] Complementing this central image in Duncan's *Venus Consolatrix* is a Celtic revival border that makes further reference to the Sphinx but also references the triple goddess of pre-Christian Celtic religion.

The Sphinx is a continuing visual theme both for Duncan and for Geddes. The implied riddle of reality is complemented in Geddes's thinking by the idea of 'the interpreter', a figure deriving not from the Greece of Sophocles or from ancient Egypt, but from the puritan thinking of John Bunyan in *Pilgrim's Progress*. This distinctly non-decadent source underpins the fact that Geddes's milieu should be primarily identified with the symbolism rather than the decadence of late nineteenth-century art. For all their similarities, a key difference between symbolist and decadent art resides in the fact that the decadent artist has no interest in spiritual transcendence, while such transcendence is fundamental to the symbolist artist.[49] It is for this reason that the work of that most transcendence-minded Puritan Christian, John Bunyan, is able to sit easily with this symbolist milieu. Reference to Bunyan is common around Geddes. For example, it is found in the title of the series of *Interpreter* magazines issued in 1896, and in the same year the Outlook Tower is referred to as 'the Interpreter's House', a key element of *Pilgrim's Progress*.[50] The first extended and systematic account of the Outlook Tower echoes this in its title; this was John Kelman's 1905 lecture, noted above. In due course Amelia Defries makes the Bunyan reference central to the title of her 1927 book, *The Interpreter: Geddes, the Man and his Gospel*.[51]

6

Manifestos in Word and Image

THE MURALS OF RAMSAY GARDEN

T HE EVENT THAT ESTABLISHED the close working relationship between
John Duncan and Patrick Geddes was the painting of a frieze for
Geddes's flat in Ramsay Garden. Duncan followed this with his most
important mural scheme, a series of large panels for the common room
of Ramsay Lodge. Duncan had returned from studying in Europe to
his native city of Dundee in 1891, and this is the time he got to know
Geddes, by then in his professorial role at University College.[1] Both
Geddes and Duncan would have been aware that an important restora-
tion of pre-Reformation mural paintings had been carried out close to
Dundee at the church of St Marnock at Foulis Easter in the late 1880s.[2]
It attracted significant local press coverage.[3] Those rare survivals of
Reformation iconoclasm would at the very least have added a sense of
historical validity to Duncan's efforts and that restoration project can be
seen as a further indication of the strength of the Arts and Crafts ethos in
Scotland.[4] It was also an indication of the fact that attitudes to religious
art had changed markedly since the Reformation, so it is something of
an irony that one of Duncan's first works for University Hall became a
late victim of iconoclasm. His painting of St Andrew was regarded as
idolatrous by the custodian of the neighbouring Free Church College,
'whose iconoclastic proclivities and horror of any representation of the
unclothed human form, prompted him, when put in charge of the building, to tear the offending canvas across the middle'.[5]

Duncan was Geddes's most enduring artistic collaborator and he
became recognised as the leading artist of the Celtic revival in Scotland.
From an early age he had contributed illustrations to the journalistic
culture of Dundee. He was, like so many Scottish artists of his day, well

travelled, having studied in Antwerp, Dusseldorf and Italy. His contact with Geddes led not only to commissions but to his appointment as director of an aspect of University Hall, the Old Edinburgh School of Art. Duncan and his small band of students, which included Helen Hay and Nellie Baxter, made a fundamental contribution to Geddes's projects. As John Morrison notes, 'Duncan was more than simply a hand giving form to Geddes's ideas.'[6] A meeting of minds would be more accurate. One can reflect on Duncan's wider intellectual network – independent of Geddes – through his friendship with William Craigie. Craigie was close in age to Duncan and both came from Dundee.[7] In due course Craigie became one of the leading academics of his generation. He was joint editor of the *Oxford English Dictionary* and founder of the *Dictionary of the Older Scottish Tongue*. Holder of the chair of Anglo-Saxon at Oxford, he also made a key contribution to lexicography at the University of Chicago. The linguistic background of his family included both Scots and Gaelic.[8] He was a classical scholar and a noted translator of Icelandic work. He had a keen interest in both Norse and Celtic mythology, contributing to Andrew Lang's series of books of fairy tales from 1894 onwards. He assisted Lang in his edition of the works of Robert Burns, and wrote his own *Primer of Burns*.[9] Thus, even before John Duncan had met Patrick Geddes, he had in William Craigie a friend who epitomised the wide-ranging academic thinking that he would find in Geddes.

In a letter dated 1893, Duncan asked Craigie's advice on details of subject matter for the mural frieze in Geddes's Ramsay Garden flat.[10] The theme was the history of pipe music and Duncan outlined the initial ideas to Craigie as proceeding from an image of reeds blown in the wind, by way of Pan teaching Apollo to play his pipes, to Hebrew pipers playing before Moses parting the Red Sea. After that Duncan proposed a Bacchanalian scene with satyrs and maenads, the Pied Piper of Hamelin and a medieval Celtic piper. The final image was to be that of the Jacobite leader Claverhouse being carried from the field of Killiecrankie in 1689 with, in Duncan's words, 'the pipes of the Highland clans that fought with him marching beside their slain chief'; indeed, he sketches this scene for Craigie.[11] That particular scene had been given visual form some years before by Joseph Noel Paton in his illustration of W. E. Aytoun's poem *The Burial March of Dundee* (Dundee and Claverhouse being one and the same).[12] Through his antiquarian interests both with respect to material culture and to literature, Paton is an important precursor of Duncan. He also contributed poetry to *The Evergreen*. Duncan's mural scheme is no longer visible but versions of three of the elements are reproduced in *The Evergreen*. These are all classical scenes: *Apollo's Schooldays* and *Pipes of Arcady*, from the *Book of Spring* published in 1895, and *Bacchus and Silenus*, published in the *Book of Autumn*, also in 1895. Indeed, writing from Kirkcudbright in July of that year, Duncan requests Geddes to return to him a watercolour sketch of his Bacchanalian procession because he is about to do his drawing for *The Evergreen* and he can't 'get along without having that sketch by me'.[13] Margaret Armour's assessment of mural painting in Scotland for *The Studio* in 1897 reproduces part of that mural,[14] while in a piece entitled 'The Decorative Paintings of John Duncan', which appeared in *The Artist* in 1898, three further elements are reproduced, namely *The Coronach of Claverhouse*, *The Pied Piper of Hamelin* and *Mons Meg*.[15] The last subject was not part of the original proposal and seems to have replaced the medieval Celtic piper mentioned by Duncan in his letter to Craigie. Since this was the scene on which Duncan asked Craigie's advice, perhaps the change was Craigie's suggestion. It has a local and international reference typical of Geddes, for Mons Meg is the famous cannon of Flemish origin sited in Edinburgh Castle only a few hundred yards from Geddes's flat in Ramsay Garden. The writer in *The Artist* notes that in Duncan's composition, which shows the cannon being drawn to the castle, pipers dressed in scarlet and green go before it and bring up the rear.[16] The entire scheme was entitled *The Evolution of Pipe Music* and one should reflect on the choice of topic as a deliberate illustration by Geddes, through Duncan's art, of an aspect of music as it relates to social evolution, from classical myth through early urban legend, to national and local historical struggle. Relevant here also is the actual culmination of that musical evolution, the ultimate set of pipes, namely the organ, for, according to *The Artist*, the frieze was in proximity to a small pipe organ positioned in Geddes's hall.[17] The whole frieze might also be read as a tacit recognition of Walter Pater's remark that 'all art constantly aspires towards the condition of music'.[18] Certainly Pater was an influential figure for the Edinburgh art world of the day. As noted in chapter three, he had in 1883 acted as a referee for John Miller Gray in his successful application for the post of curator of the Scottish National Portrait Gallery[19] and had also inspired Phoebe Traquair in her embroidery panels *The Progress of a Soul*.[20]

John Duncan remained committed in his art to what Geddes called the realities of the Scots Renascence throughout his career. Further evidence of their collaboration can be seen in a letter written in late July 1895. Duncan writes to Geddes that he has been expecting to hear from him about a symbolic design for *The Evergreen*, noting that he is ready to proceed with it whenever Geddes gives the 'ground to go upon'.[21] This is an intriguing letter, for it would seem at first sight to refer to another of the set images which includes *Lapis Philosophorum* and *Arbor Saeculorum*, both of which were by that date published (as was *Anima Celtica*, see below). Perhaps it refers to Duncan's title page design, for otherwise the nearest thing to a 'symbolic design' in the next *Evergreen*, *The Book of Autumn*, is not by Duncan at all, but *Der Zeitgeist* by Pittendrigh Macgillivray. It shares important characteristics with both *Arbor* and *Lapis* in its depiction of the Egyptian Sphinx, and extends the symbolic discourse of the Geddes circle by hinting at Rosicrucianism. The following issue, *The Book of Summer*, has no obvious image that would meet Duncan's description of a symbolic design, although the mysterious *Surface Water* is a possibility. In *The Book of Winter* a possible candidate is an image by Duncan called, simply, *The Sphinx*. This has a visual character which links it strongly to the Egyptian reference of the *Lapis* and in composition prefigures the frontispiece of Geddes's *Masque of Learning* drawn by Stanley Cursiter in 1912. Whatever the subject of Duncan's query, it reminds one of the central importance of symbolic designs for Geddes, and of Duncan's interest in giving Geddes's ideas visual form.

By April 1896 a significant part of the scheme for the Geddes family flat in Ramsay Garden had been carried out more or less as specified in John Duncan's letter to William Craigie three years before. Viewing the flat formed part of an exhibition by artists of the Old Edinburgh School of Art and a pamphlet detailing the exhibits gives unparalleled insight into its activities.[22] The committee included not only Geddes, but J. Christieson Oliphant,[23] Stewart Henbest Capper, the *Evergreen* artist James Cadenhead and Thomas B. Whitson.[24] John Duncan is listed as the director and the catalogue of the exhibition begins with his panels for the common room of Ramsay Lodge. These panels can be seen as a visual manifesto typical of Geddes's milieu. They bring together legend, history and interdisciplinary learning in a location dedicated to students. The collaborative ethos of the Old Edinburgh School of Art is immediately evident, for the works are introduced as a series of decorative pictures 'with Mural Decoration of Celtic Tracery by Helen Hay, Helen [Nellie] Baxter, and M. Mason'. All but a few fragments of these knotwork borders have been lost, but at least the main panels survive. Fortunately an illustrated account of the mural scheme was published and this shows some of the panels with their original borders.[25] Six panels are noted, all of which relate to Celtic and Scottish myth, history and educational achievement. These are, in the wording (and spelling) of the catalogue: *The Awakening of Cuchullin, The Combat of Fingal, The Taking of Excalibur, The Journey of St Mungo, The Vision of Erigena* and *Michael Scot*. A seventh panel, *The Admirable Crichton*, was completed soon afterwards, and panels commemorating Napier, Watt, Scott, Darwin and Lister were added much later, in 1926.[26] Part of the Celtic revival identity was formal, residing in the now lost interlace borders 'in the ancient Celtic taste', as Margaret Armour remarks, 'of the *Book of Kells*'.[27] The relevance of Armour's comparison can be discerned from surviving illustrations, which show a number of these intricate designs. [Figure 4.] The fact that they had more than a mere framing function for Duncan's images is made very clear in Geddes's own account of them:

> Leave the pictures, then, for a moment, and take their tracery borders. Follow them round the walls. Watch them until the interweaving threads begin to twist, and the mazy spirals spin. . . . Here are line and colour woven into patterns, changing as they go; sixteen designs round a single picture. There is no mechanical repetition, as in more familiar styles; each device is a separate living thought – a separate expression of a separate life moment of the artist, expressible only in his medium of line and colour, like a chord of music.[28]

Geddes then proceeds to find the inspiration of these designs in ripples and rainbows, and draws analogy for their enjoyment in 'the pleasure of epigram or paradox . . . for here are unity and contrast brought to satisfactory solution . . . The artist's business is to make witty inventions; ours just now to find out the knowledge of them.' And Geddes's emphasis on this education of the visual senses continues as he discusses the use of a well-decorated room 'for the convalescence of the colour sense'. He notes further that in the developing Outlook Tower, 'artist and physicist, educationist and doctor, will try to work out a simple colour sanatorium . . . beginning with soap bubbles and polariscopes, with minerals and flowers, and coming on to the decorative mysteries'.[29]

The loss of these borders, which Geddes clearly considered as fundamental to the experience of the room, is the more regrettable because these 'decorative mysteries' were the work of John Duncan's students Helen Hay, Helen (Nellie) Baxter and Marion Mason, about all of whom relatively little is known. Examples of their work can be seen in *The Evergreen* and related publications. But it should be noted that their *Evergreen* work is adapted for that black-and-white medium and as such is for the most part quite distinct from the Ramsay Lodge work, which as both Geddes and Margaret Armour note, and the surviving photographs show, is a direct homage to Celtic design of the seventh, eighth and ninth centuries. These designs must be seen not only in the context of *The Book of Durrow*, *The Lindisfarne Gospels* and *The Book of Kells*, but also with reference to the Museum of Antiquities in Edinburgh, which contained related Pictish carvings. They must also be seen in relation to one of the earliest public monuments in Celtic revival style in Edinburgh, positioned only a few yards from Ramsay Garden. It is one of the architect Rowand Anderson's earliest designs, a memorial to the 78th Highlanders, erected in 1861 on the esplanade of Edinburgh Castle. It takes the form of a Celtic cross (loosely based on St Martin's Cross on Iona) engraved with interlace. Geddes would have been well aware of not only its design but also its Gaelic inscription. The event marked by the monument was the role of the 78th Highlanders in the suppression of the Indian Mutiny in 1857–8, and in reference to that the design included both a stag's head and an elephant. The regiment was known as the Ross-shire Buffs and became a founding part of the Seaforth Highlanders in 1881. I noted in chapter two that the Black Watch's activities in India would have been salient in Geddes's child-hood. For the adult Geddes, the paradoxical relationships, evident in this monument, between imperial projects and local cultures would have been clear.

Within the interlace borders in Ramsay Lodge were John Duncan's images. The starting point for the interpretation of these images is set by an illuminated quote, 'As it hath been so it shall be'. This is a contraction of a passage from the book of Ecclesiastes chapter one, verse nineteen, which reads in the Authorised King James Version of the Bible: 'The thing that hath been, it is that which shall be . . .'. The quotation appears over the doorway and these words (and indeed Ecclesiastes as a whole) resonate not only with the cyclic aspects of Geddes's thinking, but with a poem attributed to St Columba which concludes: 'But ere the world come to an end / Iona shall be as it was'. The relevance of that poem lies in the fact that it was quoted by Geddes in his *Evergreen* essay *The Scots Renascence* published in 1895. Columba was a figure of great interest to Geddes's circle, the more so because the 1300th anniversary of his death fell in 1897.[30] The Ramsay Lodge quotation has a clear association with those by Burns and Aeschylus already discussed with respect to the Ramsay Garden sundial, but where those writers are concerned with time as it relates to change, both personal and histori-cal, the words in the Ramsay Lodge mural are concerned with time in an explicitly revivalist manner. That revivalist quality is underpinned by the image that the quotation surrounds. This is John Duncan's *The Awakening of Cuchullin*, which shows the Celtic hero awakening after a restorative sleep, his wounds healed.[31] This is the anchor image of the series and leads on to *The Combat of Fingal*,[32] which shows a scene derived from Macpherson's *Ossian*. The third panel moves us from Gaelic Celticism to the Brythonic Celticism of King Arthur in *The Taking of Excalibur*. The next continues that southern Scottish Brythonic theme with the image of *The Journey of St Mungo*, at the same time introducing historical Christianity into this visual history of Scottish ideas and legends. Following it is an image inspired by the writings of the great Gaelic-speaking Irish theologian of the ninth century, *The Vision of Johannes Scotus Erigina*. That dream is complemented in the next panel by the thinking of the mage and early scientist of European reputation, *Michael Scott*, renowned for his translations of Aristotle from Arabic. These latter two figures show a significant transition in the series, for they indicate the beginning, in the medieval period, of an intellectual tradition continuous with the present. The final figure of the first set (not painted in time for the 1896 exhibition) was *The Admirable Crichton*. Crichton, shown by Duncan during his visit to Venice which preceded his death in a duel in Mantua, was the Scottish and European Renaissance scholar *par excellence*. His inclusion strikes a personal note for Geddes, for Crichton is thought to have been educated at Geddes's old school, the Grammar School of Perth, before studying at St Andrews University. Duncan was still waiting for payment for this last panel in 1898, as a letter to Geddes shows.[33] The letter also sheds light on the wider Scottish arts and crafts links of the time, for in it Duncan mentions that he has been invited to become a founder member of 'an arts and crafts society forming in Glasgow' and his pride in this association of artists is reflected by the fact that he lists all nineteen other founder members. It is an impressive list and includes Francis and Jessie Newbery, Phoebe Traquair, Pittendrigh MacGillivray, David Gauld, Charles Rennie Mackintosh and Margaret Macdonald. Duncan also notes in this letter that he has enlisted as an assistant in Dundee 'a very promising young artist', a reference to the talented but short-lived George Dutch Davidson.[34] Davidson's work – which includes an illustration derived from Fiona Macleod's *The Hills of Dream*,[35] made in 1899 – forms a stylistic bridge between Duncan and the Glasgow artists associated with Mackintosh.

Even before the *Admirable Crichton* panel had been completed, an explanatory document had been written by J. Arthur Thomson for Geddes's series of pedagogical pamphlets, *The Interpreter*.[36] An illustrated version of this pamphlet was issued in due course, and this continued to be reprinted under the imprint of University Hall at least as late as 1944. A further example of a Geddes manifesto in mural form can be found in Duncan's painting *Joan of Arc and her Scots Guard*.[37] This work draws attention to the international links between France and Scotland and was originally located in the common room of the University Hall residence of St Giles House, a few hundred yards east of Ramsay Lodge. The Auld Alliance of Scotland and France was a subject of enduring interest to Geddes and the subject of Joan of Arc was first mooted by Geddes in a letter to Duncan sent in November 1895. In it he notes a suggestion from Andrew Lang for a painting of Joan of Arc with her bodyguard of Scottish Archers.[38] Geddes describes Lang as 'of the Franco-Scottish society', which was founded around this time. In due course a black-and-white version of the design appeared in *The Evergreen* in *The Book of Summer* in 1896 under the title *The Way to Rheims*. It is captioned 'Franco-Scottish Society – Sorbonne, 18th April 1896' in honour of the first meeting of the society on French soil. Andrew Lang was at the same time engaged in another project related to Joan of Arc, his novel *A Monk of Fife*, which tells the tale of a Scottish cleric from St Andrews in the France of St Joan.[39]

THE EVERGREEN: A NORTHERN SEASONAL

Another intriguing visual manifesto associated with Geddes can be found in *The Evergreen*. This is John Duncan's *Anima Celtica*, a drawing published in 1895 in *The Book of Spring*. It is a remarkable synthesis of Celtic and related symbolism: standing stones, the blind Ossian accompanied by Malvina, Fingal with the newborn Ossian, the awakening of Cuchullin, an ancient bronze sword of Celtic pattern, all centred on a Celtic muse, the soul of the Celt herself. A further aspect is Jacobite symbolism, which can be seen in the foreground and takes the form of an eighteenth-century dirk and a bonnet with a white cockade. The message is clear, namely that there is continuity between the Celtic revival of the late nineteenth century and the Jacobite period. But the old Celtic sword extends this continuity to periods before written history. The wider message is thus that the Celtic revival is part of a continuous Highland tradition, which leads back, via James Macpherson, via the Jacobites, to the time of the Fenians and before.[40]

In *The Evergreen* the manifesto quality of Duncan's drawings was complemented by Geddes in a set of six essays (one co-authored with J. Arthur Thomson) in which he set out his cultural and ecological thinking. Taken together these essays demonstrate Geddes's interdisciplinary endeavour. In order of publication, the first of these was *Life and its Science*, an essay dealing with ecology and evolution and joined in the same issue of *The Evergreen* by Geddes's key statement on cultural revival, *The Scots Renascence*. In *The Evergreen's* second issue, *The Book of Autumn*, Geddes wrote a 'prefatory note' with Victor Branford which outlined the cultural revivalist aims of the magazine; in the same issue the relationships of art and science were explored in his essay *The Sociology of Autumn*. In the next issue, *The Book of Summer*, Geddes contributed *The Flower of the Grass*, a meditation on anthropology and the development of civilisation, which reflects in its title his commitment to the language of the Bible. Its Latin translation, *Flos Herbae*, gave its title to one of Geddes's stained-glass teaching windows for the Outlook Tower. *The Book of Summer* also contained *The Moral Evolution of Sex*, written jointly with J. Arthur Thomson and extending further into the social realm of Geddes and Thomson's popular book, *The Evolution of Sex*. In *The Book of Winter*, the final *Evergreen*, Geddes explored city planning with reference to its pre-historic origins in *The Megalithic Builders*, and wrote the concluding 'Envoi' with William Macdonald. While these essays are clearly of their time, the illumination provided is often as relevant today as it was then. They are linked by a commitment to evolution, biological, social and cultural, and an energetic refusal to observe disciplinary boundaries. There is also a powerful advocacy of an integration of ecology and religion which looks forward to the social movements of the 1960s and 1990s. For example, in *The Moral Evolution of Sex* there is a paragraph which holds that romance and poetry are truly religious: 'For religion lies in idealising and consecrating life; and love is life, and life is love; so Robert Burns, human sinner, is also sacred bard.' Geddes (or Thomson) continues, 'The Nature-Religions, like all others, are not dead, but are returning; and in ever purer forms.'[41]

Around these essays, which one can regard as the ideological backbone of *The Evergreen*, are others by – among others – Élisée Reclus, Charles Sarolea, J. Arthur Thomson and Alexander Carmichael. These are complemented by the poetry and prose of William Sharp (under his own name and as Fiona Macleod), Douglas Hyde, Pittendrigh Macgillivray, Edith Wingate Rinder, Joseph Noel Paton, Margaret Armour and others.[42] I have noted that several of the images, like *Arbor Saeculorum*, *Anima Celtica* and *The Way to Rheims*, have a polemical and a pedagogical quality. Another that must be mentioned from that perspective is *Antarctic Summer* by W. G. Burn Murdoch. It shows penguins on an ice sheet and is without doubt the most unexpected image in *The Evergreen*, but its appearance firmly links the notion of global geographical awareness to cultural revival. The image is based on observations made by the artist when he accompanied his fellow Geddes student William Spiers Bruce to the Antarctic in 1892–3. The invitation came from Bruce during lunch at University Hall in Edinburgh and a week later they left on a whaler from Dundee, the *Balaena*.[43] The following year Geddes and his wife met them on the quayside on their return to Camperdown Dock.[44] Bruce was frustrated by the commercial aims of the expedition, but Peter Speak has pointed out that despite that limitation 'virtually no scientific work had been undertaken in the Antarctic between the James Clark Ross expedition with the *Erebus* and *Terror* (1839–43) and the Dundee whaling expedition', and that consequently Bruce, as scientific officer, had restarted the international scientific study of Antarctica after a break of half a century.[45] From the perspective of cultural revival, the following comment from Burn Murdoch is germane. He recalls that prior to Bruce's subsequent Antarctic endeavour, the Scottish National Antarctic Expedition, he told Burn Murdoch 'in an unusual burst of confidence, that the expedition was really the outcome of Geddes's influence, meaning that it was Bruce's contribution to the Scots renascence'.[46] Nothing could more clearly illustrate Geddes's ethos of cultural revival depending on an interdisciplinary approach. The writings and lectures given by Bruce and Burn Murdoch on their return in 1893 'started an interest and led to Bruce's further and purely scientific voyages which placed him in the forefront of naturalist navigators and explorers'.[47] Burn Murdoch wrote extensively and in evocative terms of the 1892–3 expedition, recalling: 'icebergs a thousand feet thick and miles and miles long, crushing together or ploughing through smaller bergs and floe ice 20 feet thick, piling it into packs of 50 to 80 feet in almost the twinkling of an eye, into wonders of cyclopean ice architecture; then in a shift of current or wind all clear and open sea for miles!'.[48]

Burn Murdoch notes of his student days in the hall of residence at Riddle's Court that Bruce and himself, though they had 'gained little materially out of any Geddesian scheme', were 'deeply indebted to his genius, and to the friends he collected round him, for raising our spirits a little out of the rut into what we may call the dreamland of science'.[49] Burn Murdoch tells not only of the influence of Geddes on Bruce, but also of the influence of Sir John Murray, J. Arthur Thomson and others. W. S. Bruce's biographer illuminates this further: 'among that little coterie in Geddes's halls, who had a common belief in the possibility of a Scots renascence'. He continues: 'That spirit found expression in many ways; the periodical called the *Evergreen*, the Celtic verse and stories published by Geddes, the paintings of Burn Murdoch, John Duncan and others, the Outlook Tower and the buildings of the halls themselves.

All were striving to awaken the dormant Celtic spirit, to rekindle the national fire and to show what Scotland could do in various walks of life.'[50]

The 1896 exhibition catalogue moves on with works exhibited in another part of Ramsay Lodge, Allan Ramsay's Drawing-Room. Here another key artist of Geddes's circle is introduced, Charles Mackie. He is represented by two seasonal murals, an *Autumn Pastoral* and a *Spring Pastoral*. Mackie is a most interesting figure, not least because of his close links with Gauguin and Sérusier. He was a visitor to Pont-Aven and the influence of synthetism is clear in his work. Mackie was also a major contributor to *The Evergreen*, giving the magazine its initial visual identity through a set of cover designs using a tree of life motif. Along with Mackie's murals, the Drawing-Room also contained originals of drawings reproduced in *The Evergreen* by Hay, Cadenhead, Duncan and A. G. Sinclair: Helen Hay's almanacs were certainly on show and Duncan's *Anima Celtica* was presumably there also. Had the exhibition been of wider focus than the Old Edinburgh School of Art a work by Paul Sérusier, *Pastorale Bretonne*, might have been included, for Mackie's contact with Sérusier had led to the latter contributing this image to the first issue of *The Evergreen*. On the basis of correspondence between the two, Claire Willsdon has raised the possibility that Sérusier painted a mural for Ramsay Garden.[51] There is no independent evidence of this but what is certain is that a mural project was discussed and Sérusier's work is linked in theme and format to Mackie's murals for Geddes's flat. The association between Mackie and Sérusier emphasises that the Breton primitivism of the Pont-Aven School was an aspect of the pan-Celticism of the late nineteenth century.[52] More widely, Willsdon has noted Duncan and Mackie's affinities 'with Continental Symbolism's ideology of mural painting as an affective medium affording entry into dreams and emotional states', going on to note Duncan's reading of the theories of Sar Péledan, as well as those of Burne-Jones.[53]

Also in this part of the exhibition were a number of other works by John Duncan, including *The Witches Well*, cast in bronze by D. W. Stevenson. This was in due course positioned close to Ramsay Garden, fixed to the wall of Castlehill reservoir near the spot on the Castle esplanade where witches had been burned. Castlehill reservoir was at that time an enclosed header tank for the city linked to the reservoirs of the Pentland hills. Typical of a Geddes project, a memorial to the persecution of knowledge-bearers (this knowledge often, as the design of *The Witches Well* makes clear, of a botanical and medicinal nature) was juxtaposed with another type of source, the geological actuality of a water supply. The next room contained a substantial amount of work by James Cadenhead and also further work by Duncan, which included two watercolours from a series illustrating the classical myth of Orpheus. These latter relate to a mural project carried out in 1897 at Pitreavie Castle near Dunfermline for the textile manufacturer Henry Beveridge. Duncan's *Orpheus* owes a stylistic debt to Phoebe Traquair's Pater-inspired work-in-progress, *The Progress of a Soul*, referred to above. His immediate literary source for this Greek myth is the Scots-language poem 'Orpheus and Eurydice' by the fifteenth-century Dunfermline poet Robert Henryson. Thus this series was again both international and local in reference. The *Studio* article of 1897 also illustrates several designs intended for Pitreavie by Charles Mackie. These have a particular interest because just as Duncan's *Orpheus* works relate to fifteenth-century Scottish poetry, so Mackie's relate to old Scottish ballads, in particular 'Sir Patrick Spens' and 'Hardyknute', the latter ballad at the time attributed to a one-time owner of Pitreavie Castle, Lady Wardlaw.[54] A further layer of this cultural project which links it directly to Ramsay Lodge and University Hall is that 'Hardyknute' and 'Sir Patrick Spens' had been collected by Allan Ramsay in the early eighteenth century for *The Ever Green* and *The Tea-Table Miscellany*. This has a further significance when seen in the context of Geddes's *Evergreen*, for poetry in Scots is a key part of the cultural revivalist mix of Geddes's project.

Early indications of this are found in *The New Evergreen* in a poem, 'The Skipper's Bride', by Gabriel Setoun; and in *The Evergreen: Book of Spring*, Setoun contributed another poem in Scots. Helen Hay makes an interesting contribution here as editor of an undated pamphlet entitled *For Auld Lang Syne*, which very possibly marked the Burns centenary of 1896. It is certainly closely related (both by typography and by the printer, Riverside Press) to the publications of Geddes's Celtic Library at that time. The cover may have been the work of Nellie Baxter, for there is an 'Auld Lang Syne' cover design by her noted in the Old Edinburgh School of Art exhibition catalogue. The pamphlet contains poems by various hands, linked by the theme of Edinburgh. Crucial to the selection are Scots-language works by both Ramsay and Burns. Another *Evergreen* contributor who wrote in Scots for the magazine, as well as contributing images, was Pittendrigh Macgillivray, by this time at the height of his powers as a sculptor but also a poet of no mean ability.[55] An imitation of an old ballad, 'Gledha's Wooing', was contributed by John Geddie. The following year, 1896, Geddie's *The Balladists* was published as the sixth volume of Oliphant, Anderson and Ferrier's *Famous Scots* series.[56] This series was yet another interdisciplinary cultural nationalist publishing project of the time. The five volumes which preceded Geddie's book had been devoted to Thomas Carlyle, Allan Ramsay, Hugh Miller,[57] John Knox and Robert Burns, and one finds here a kind of literary equivalent of the Scottish National Portrait Gallery.[58] In *The Balladists* John Geddie provides background for Geddes's agenda for *The Evergreen* when he describes the importance of the old ballads collected in Allan Ramsay's *Ever Green* and *The Tea-Table Miscellany* as 'a fresh dawning of Scottish poetry . . . warmth, light, and freedom seemed to come again into the frozen world'.[59] In *The Book of Autumn*, the second issue of *The Evergreen*, Geddes and Victor Branford echo these sentiments and take them into their present: 'In 1724 Alan Ramsay published his "Evergreen", desiring thereby to stimulate the return to local and national tradition and living nature.'[60] In *The Evergreen* reference to Ramsay was, not surprisingly, visual as well as verbal; for example, Charles Mackie's *Robene and Makyn*[61] illustrates a ballad by Henryson collected by Ramsay in *The Ever Green*. Mackie's *Robene and Makyn* and Duncan's *Apollo's Schooldays* both served to illustrate a review of *The Evergreen: Book of Spring*[62] which appeared in *The Artist* in July 1895. As in the case of *The New Evergreen*, the review was very positive. The final *Evergreen*, the *Book of Winter*, again attracted praise and the hope that *The Evergreen* heralds 'a long series of similar publications'.[63] Again there was a good-sized illustration, *The Victor* by Robert Burns (the artist, not the poet). *The Evergreen* was not reviewed in *The Studio* until 1897, when *The Book of Winter* was the subject. Geddes had conceived the magazine as a 'seasonal' which would have one issue for each of the four seasons. It is often mistaken for a quarterly, but in fact only two issues were produced each year. The first and only cycle was completed after two years, although a second cycle was mooted. *The Studio* commented that the cessation of 'this very original periodical' after its first season-cycle is 'altogether to be regretted' and the hope was that *The Evergreen* would 'put forth fresh leaves and buds and blossoms'. The review was given visual emphasis by a full-page reproduction of a landscape by James Cadenhead. The author of the review is not given but by this time *The Evergreen* had already been noted by Walter Crane in his canonical work *Of The Decorative Illustration of Books*. Crane wrote there that the 'publication of *The Evergreen* by Patrick Geddes and his colleagues at Edinburgh has introduced several black and white designers of force and character' and goes on to note Burns and Duncan as 'distinguishing themselves for decorative treatment in which one may see the influences of much fresh inspiration from Nature'.[64] This assessment is backed up by three full-page illustrations, two of work by Duncan, one by Burns.[65] It should be noted that Crane thought highly enough of these artists, and was aware enough of their activities, to include work by them published only the year before his own book. One of the images by Duncan, *Apollo's Schooldays*, is intriguing for it appears in Crane's book in a slightly different version from its appearance in *The Evergreen*. The two images are identical except for the addition, in the Crane publication, of a dotted outline representing Apollo's shoulder. The other difference is that the block-maker, Hare, has signed himself on the left-hand side of the print, whereas in the *Evergreen* version his signature appears on the other side. One can speculate that Crane is using a trial proof from which to reproduce. Why that is the case is not clear; perhaps he was able to obtain proofs prior to publication of *The Evergreen*. As a letter of congratulation to Geddes shows, he certainly had access to the *Book of Spring* soon after its publication.[66] Further Scottish interest comes in the inclusion in Crane's book of Selwyn Image's title page decoration for the by then defunct *Scottish Art Review*.[67] Notable also is Crane's mention of the high quality of printing available in Edinburgh from both R. and R. Clark and T. and A. Constable.[68]

Further insight into the developing milieu of Celtic revival artists of the period is given in a letter from Duncan to Geddes. Writing in July 1895, he requests two copies of the first issue of *The Evergreen*, one for himself and one for a fellow Dundee artist, Stewart Carmichael.[69] Carmichael is an interesting figure in his own right. He contributed in due course in 1903 to a Dundee student magazine, *The Meal Poke*, which was clearly inspired in its design by *The Evergreen*. Two of the editors are Geddes's collaborators R. C. Buist and H. Bellyse Baildon, although there is no direct reference to Geddes and no participation by John Duncan. As late as 1930 one finds Stewart Carmichael illustrating a Celtic revival magazine which is in format, like *The Meal Poke*, a direct descendant of *The Evergreen*. That publication, *An Rosarnach* (the rose-garden), is wholly in Gaelic.[70]

The next section of the Old Edinburgh School of Art exhibition catalogue is an invitation to visit the Geddes flat every afternoon for the duration of the exhibition to see not only the murals by Duncan but other parts of this 'decoration of a private house', as it is described, including work by Charles Mackie, Mary Hill Burton and Helen Hay. Most of this work has been destroyed or is not visible but four notable panels by Mackie, each showing a season, were discovered in the 1990s concealed under layers of paint. Mary Hill Burton painted murals not only in Geddes's flat but also in the University Hall residence in St Giles

Street, and that scheme attracted notice in *The Studio* in 1898.[71] The final part of the Old Edinburgh School of Art exhibition took place in the studio of the school at 7 Ramsay Lane. Gathered there were all sorts of products of applied art with Celtic decoration. Included were accessories of Highland dress, notably work by John G. Milne, for example a 'Skene Dhu, with blade embossed with Celtic ornament', a dirk, four quaichs, and a sporran 'with brass top engraved in Celtic twists'. At around the same time, Alexander and Euphemia Ritchie of Iona were developing similar work for sale. Their Celtic-patterned work in brass, silver, pewter and leather is now sought after.[72]

Such articles can be seen in terms of the continuities of tradition noted above with reference to John Duncan's *Anima Celtica*. As I have noted, these objects included a Highland dirk of eighteenth-century pattern and a Celtic sword. Both weapons, and by extension the exhibits in that part of the exhibition, are to a significant degree dependent for their visual reference on the activities of the Society of Antiquaries of Scotland.[73] The museum of the society occupied half of the new Scottish National Portrait Gallery building, and both the architect of the building, Robert Rowand Anderson, and the funder of the building, John Ritchie Findlay, as well as being Edinburgh Social Union members, were fellows of the society. Other fellows of the society who had immediate links to Duncan or to Geddes include William Craigie, Henry Beveridge, Gerard Baldwin Brown and W. B. Blaikie. The exchange of knowledge here can be imagined. For example, in 1892, the year Duncan's friend Craigie was elected a fellow, an article in the *Proceedings* of the society by Joseph Anderson recorded the find of a leaf-bladed sword by a crofter cutting peat at Aird in Lewis.[74] One of the unusual aspects of this find was the good state of preservation of the horn handle, and it is possible that Duncan was influenced by this find in the design of the handle of the sword in *Anima Celtica*. Duncan's version is by no means an exact copy, but my concern here is to indicate the type of visual material that would have been available to Duncan, rather than to suggest an exact relationship.[75]

Hugh Cheape has noted with respect to the antiquarian aspect of Duncan's work that 'it is no doubt significant that the National Museum of Antiquities was reorganised in new premises in Queen Street in 1891', that is to say in the eastern part of the Scottish National Portrait Gallery building.[76] This move was accompanied by the publication of a well-illustrated catalogue. As well as the weapons noted above, Duncan used material such as penannular brooches to give authenticity to his Celtic subjects. Such brooches can be seen, for example, in *The Awakening of Cuchullin* for Ramsay Lodge, in *Anima Celtica* for *The Evergreen* and, much later, in *Cuchulain*,[77] a drawing dating from about 1912 which was reproduced in another Geddes magazine project, his civic review *The Blue Blanket*. That drawing was later used as the frontispiece for the second volume of *Songs of the Hebrides* collected and arranged by Marjory Kennedy-Fraser and Kenneth Macleod. Duncan's use of standing stones in *Anima Celtica* is also resonant with the material published by the society, but has an even closer link to two essays in *The Evergreen*, Geddes's *The Megalithic Builders* and George Eyre-Todd's *Night in Arran*. The former has already been noted; the latter, which appeared in *The Book of Summer*, is an imaginative re-creation of the rituals of a pre-historic Goddess religion among the stone circles of Machrie Moor on the island of Arran. William Hole's mural *The Mission of St Columba to the Picts*, completed in the Scottish National Portrait Gallery in 1898, bears interesting comparison with the antiquarian aspects of Duncan's *Anima Celtica*, for he too depicts Celtic bronze weaponry and standing stones, not to mention drawing on the form of a king from the Lewis chess pieces of the Norse period to provide the model for his Pictish king. It would be easy to dismiss the confusion of the chronology in Hole's work as some sort of anachronistic decorative licence, but that would be to miss the point. The purpose of Hole's work is to draw attention to objects in the collection of the Museum of Antiquities, then sited only a few yards from his painting, rather than to create an accurate historical image.[78] This pedagogic intention is underlined by Hole's use of some of the same objects, for example the labyrinth-like, bronze shield found at Auchmaliddie, in his frieze of figures from Scottish history in the same building. Close by he asserts in visual terms the twin currents of Highland Gaelic and Lowland Scots identity in panels entitled *Pibroch* and *The Ballad*. Hole began his portrait gallery work in 1897, so would probably have been aware of John Duncan's mural scheme on the evolution of pipe music painted for Geddes's flat. He might well have included the subject of pibroch anyway, but it is at the very least an interesting resonance between the two projects. However, as Duncan Comrie has pointed out, in comparison to Duncan's work for Geddes, 'explicit Celtic cultural identity' was distinctly muted in the Portrait Gallery scheme. This is resonant with notable omissions in the Portrait Gallery frieze. The two great Gaelic poets of the eighteenth century, Duncan Ban MacIntyre and Alexander Macdonald (Alasdair mac Mhaighstir Alasdair), simply do not figure. And there is also no place for another eighteenth-century Highlander, one of the most internationally influential of all Scottish writers, James Macpherson.[79] The omission of Macpherson betrays a reading of Scottish literary history through the prejudices of the Londoner Samuel Johnson, rather than through the independent judgement that one might have hoped for from those who specified the figures to be represented. By contrast, Patrick Geddes and John Duncan, both in the Ramsay Lodge murals and in *The Evergreen*, recognised in Macpherson's work the first stirrings of the Celtic revival in which they were themselves engaged. As is so often the case with Geddes, his view has stood the test of time. Thus, while Macpherson was being excluded from academic consideration, Geddes and his colleagues were issuing a new edition of *The Poems of Ossian*, edited by William Sharp. This was one of the book designs exhibited at the Old Edinburgh School of Art in April 1896, together with related book cover designs and decorations. Included were John Duncan's Celtic designs for the covers of *The Centenary Burns*. This four-volume set published by T. C. and E. C. Jack was edited by W. E. Henley and T. F. Henderson and contained etched illustrations by William Hole. It was one of many works and events that marked the centenary of Burns's death in 1896. One might have expected a more active involvement with the Burns celebrations from Geddes, but the reason for his relatively muted approach is clear when one realises the extraordinarily high level of Burns-related activity around the centenary year.[80] If there was one cultural figure who did not require revival it was Burns, and Geddes was never one to devote energy where it was not needed. Geddes's personal regard for Burns is made clear in his writings,[81] but he may have found the widespread adulation of the poet too critically unfocused for his cultural purposes, rather as Hugh MacDiarmid was to find in the 1920s.[82] What Geddes and his milieu did was to extend the boundaries of the discourse and give Burns's works a much wider cultural context within which to be appreciated. Thus, while John Duncan may have designed the cover of *The Centenary Burns*, the focus of Geddes's centenary publishing in 1896 was not Burns but Ossian, for that year also marked a century since the death of James Macpherson. The new edition of *The Poems of Ossian* formed part of the Celtic Library published by Patrick Geddes and Colleagues. It is to a consideration of that publishing venture, and the insight it provides into the cultural politics of the Celtic revival of the 1890s, that I now turn.

7

Models of Celticism

THE CELTIC LIBRARY

THE PUBLISHING OF THE centenary edition of *The Poems of Ossian* by
Patrick Geddes and Colleagues in 1896 was a timely reminder that
the roots of the Celtic revival of the 1890s lay well over a century earlier
in the work of James Macpherson. Originally published in the 1760s,
Ossian was Macpherson's response to the oppression of his own Gaelic
culture that followed the Battle of Culloden in 1746. One can illustrate
that situation by noting that in 1762, the date of publication of *Fingal* (the
first full volume of *Ossian*), it would have been illegal for Macpherson
to wear tartan, and that ban would remain in force for another twenty
years.[1] Such measures applied to all Highlanders, regardless of which
side they had been on at Culloden. It was government-sponsored cultural destruction, pure and simple. The significance of Macpherson's
work was that it challenged that destruction by making Celtic culture
of popular interest throughout Europe and North America. It is an
irony that despite numerous editions, Macpherson's *Ossian* was then
itself suppressed by being excised from the canon of English literature.
An interesting example from Geddes's time of the uncritical acceptance
of that excision was the pronouncement by Henry Grey Graham in
Scottish Men of Letters in the Eighteenth Century, published in 1901,
that 'the poems of Ossian are unread today, and will seldom be read
again'.[2] That comment is all the less perceptive when one bears in mind
that the 1896 Geddes publication of *Ossian*, edited and introduced by
William Sharp, was a pointer towards the proper scholarly assessment
of Macpherson. That scholarly assessment would only begin in earnest
when Derick Thomson's *Gaelic Sources of Macpherson's Ossian* was
published over half a century later.[3]

The centenary edition of *The Poems of Ossian* was beautifully pro-
duced. In readability it recalls the clarity of the editions from the 1760s.
As editor of the Canterbury Poets series for the publisher Walter Scott,
William Sharp would have been well aware of the pocket edition of
Ossian published in that series in 1888 with an introduction by George
Eyre-Todd. Sharp's own edition for Geddes moves *Ossian* back from the
realm of the pocket edition (which had begun a century earlier with the
Cameron and Murdoch edition of 1796) into the realm of the library.
This is an important shift. Regardless of the academic quality of Sharp's
edition (and it has its detractors), I would argue that it was through
this edition that *Ossian* became again a significant part of academic
culture. The endpapers and title page decoration of the centenary *Ossian*
were designed by Helen Hay. It was a key part of the Celtic Library,
a series published by Geddes with William Sharp's advice. Sharp first
met Geddes in the autumn of 1894. Thus began a friendship which
had far-reaching results, not least for Sharp in his intriguing *alter ego*
of Fiona Macleod. For Sharp's wife and biographer Elizabeth, both her
husband and Geddes were 'idealists, keen students of life and nature;
cosmopolitan in outlook and interest', and at the same time ardent Celts
who were 'eager to find some way of collaboration'.[4] William Sharp
became the first manager of the publishing firm of Patrick Geddes and
Colleagues, but he speedily transferred to the more appropriate role of
literary advisor. A number of Sharp's Fiona Macleod works were published by Geddes. The 1896 exhibition included cover designs for two
of these, *The Sin-Eater* and *The Washer of the Ford*. Along with them
were *The Fiddler of Carne* by Sharp's friend, the Welsh-English writer
Ernest Rhys,[5] and *The Shadow of Arvor*, a book of Breton legends done
into English by another friend, Edith Wingate Rinder, who seems to
have been Sharp's inspiration for Fiona Macleod.[6] Helen Hay designed
the covers, endpapers and title page decoration, and her designs were
complemented by the binding of William Wilson and the high production values of the Riverside Press in Edinburgh. The Celtic Library set a
standard for relatively inexpensive books of high quality that would be
emulated by T. N. Foulis ten years later.

The most intriguing of all the Celtic Library books was *Lyra Celtica*,
'an anthology of representative Celtic poetry' edited by Elizabeth Sharp.
It is a substantial work and its importance was later recognised by the
influential Irish Celticist Alfred Percival Graves in his *Book of Irish
Poetry*. He called it 'that beautiful volume, *Lyra Celtica*, selected with
great distinction'.[7] In her biography of her husband, Elizabeth glosses
over her own contribution to that major anthology. That she was a

talented editor has already been noted in chapter three with respect to *Women's Voices*. In that earlier work her editing had been complemented by the designs of Phoebe Traquair. She was equally fortunate in having Helen Hay as her collaborator on *Lyra Celtica*. The book had an introduction and notes by her husband, and its contents are worth reflecting on in some detail. William Sharp's introduction extends to over thirty pages and includes poems not in the main anthology, among them Moira O'Neill's 'Sea Wrack', Robert Louis Stevenson's 'In the Highlands' and Rosa Mulholland's 'Under a Purple Cloud'. The last mentioned had just appeared in *The Evergreen: Book of Autumn*. Also quoted at length in the introduction is W. B. Yeats's 'Wanderings of Oisin'. The anthology itself contained almost two hundred individual works, followed by forty-seven pages of notes. Contributions included ancient and modern work by poets from Scotland, Ireland, Brittany, Cornwall, Wales and the Isle of Man. *Lyra Celtica* was emphatically pan-Celtic in its inspiration, ranging in reference from Macpherson's *Ossian* to Yeats's 'The Lake Isle of Innisfree'[8] via Duncan Ban MacIntyre, Alexander MacDonald (Alasdair mac Mhaighstir Alasdair), Douglas Hyde, Sir Samuel Ferguson, Kuno Meyer, John Stuart Blackie, Hall Caine, T. W. Rolleston and Fiona Macleod. It was, however, a determinedly English-language presentation of Celtic poetry, and for William Sharp that was part of an agenda to claim the future of Celticism for English speakers like himself. To aid the idea of the renaissance of the Celt depending on an Anglophone guise, at the conclusion of his introduction Sharp quotes himself through the words of his *alter ego* Fiona Macleod: 'The Celt falls, but his spirit rises in the heart and the brain of the Anglo-Celtic peoples, with whom are the destinies of the generations to come.'[9] As a successful English-language writer, that demise of the Celt to be reborn as an English speaker was very much in Sharp's interest, and he notes W. B. Yeats as the leading poet of this new wave of English-language Celticism.[10] Yeats nearly became an author in the Celtic Library (and, indeed, in *The Evergreen*) but, presumably due to the number of books being produced in 1896 and the concomitant financial pressures, Geddes had turned down a book proposal from the Irish poet. In a letter to Sharp of 30 June 1896, Geddes wrote: 'I have written to Mr Yeats to tell him that while we are in complete sympathy with him in his project, we cannot take up his book'.[11]

WILLIAM SHARP AND ALEXANDER CARMICHAEL

While Sharp's English-language Celticism was dominant in the Geddes circle, it was by no means wholly characteristic of it. Geddes was well aware of Gaelic as his father's native language and, in the context of her husband's work with Geddes, Elizabeth Sharp mentions the presence in Edinburgh of one of the key Gaelic speakers of the Celtic revival, Alexander Carmichael. In the first volume of *The Evergreen*, Carmichael made a direct contribution to the synthesis of Scots-language and Celtic revivals by speculating on the Celtic, indeed bardic, aspects of Robert Burns's ancestry.[12] Elsewhere he offers a quite different take on Celtic revival to that of William Sharp. Carmichael was an outstanding collector and translator of Gaelic oral material. Unlike Sharp, he was a native Gaelic speaker and was recognised by Highland communities as one of their own, an attribute that Sharp was only able to claim for Fiona Macleod because she did not exist. Carmichael's most important work, *Carmina Gadelica/Ortha Nan Gàidheal*, was published four years after *Lyra Celtica*. William Sharp (in the person of Fiona Macleod) reviewed it in *The Nineteenth Century* in 1900, and it is interesting to see how he insists that Carmichael's work is a wonderful elegy for a more or less dead culture, rather than an indication of any potential for regeneration. He writes of 'the bitter solace of absorption in the language, the written thought, the active, omnipresent, and variegated energy of the dominant race', and sees nothing but an ever more insistent silence for the Gael.[13] He drives the point home with references to twilight and winter. But this is not one of Geddes's winters that leads inevitably to spring. Sharp's winter implies only death, and cultural revival is only possible through discovery by others and advocacy in another language. In short, the Celt must become like the ancient Greek: dead but admired. For Sharp, Carmichael can do no more than 'preserve awhile' a 'sunset beauty'. Having said that, Sharp does not entirely misrepresent Carmichael here, for Carmichael himself saw his work as very much a reflection of an already passed age, 'a stone upon the cairn of those who composed and those who transmitted the work'.[14] But the difference lies in Sharp's emphasis time and again on a single outcome, the death of the language. For Sharp the only option was to preserve what could be preserved of the Celt by adopting the dominant language of English. Sharp's biographer Flavia Alaya has suggested that this reflected in Sharp a pragmatic cosmopolitanism.[15] That view is not without merit, particularly when seen in the context of Sharp's response to the use of French by the Flemish writer Maurice Maeterlinck.[16] However, a less charitable critic might see

Sharp's position as representing an insidious linguistic imperialism, or at best an indication of a selective ignorance of that in which he claimed to be expert. From a psychological perspective it can be noted that Sharp's mouthpiece for most of his commentaries on matters Celtic was not simply a pseudonym, but a person who was held to exist, but who did not. Sharp's pronouncements on Celtic culture were part and parcel of that fantasy. This might seem a rather negative assessment of Sharp, but paradoxically it allows one to give Fiona Macleod's works much greater value than is often ascribed to them; indeed, it allows one to situate them on a line of development of fantasy literature that stretches from Macpherson's transformations of Celtic story in *Ossian*, via George MacDonald to C. S. Lewis and J. R. R. Tolkien. What is important with respect to my argument here is that Sharp's pronouncements on Celtic culture, whether issued under his own name or as Fiona Macleod, should be seen in the first instance as part of that fantasy world. Thus one must distinguish Sharp's yearning for the death of Celtic culture from Carmichael's matter-of-fact recognition of loss.

Alexander Carmichael's *Carmina Gadelica* is a remarkable document. It is a collection of hymns and incantations in Gaelic and in English, and it is the work not just of a great collector but of a great translator. As John MacInnes notes, Carmichael's editing of his orally collected material has been the subject of some controversy.[17] But what has never been in doubt is the linguistic quality of the words, both Gaelic and English, he brings to the reader; thus MacInnes quotes with approval Ronald Black's description of *Carmina Gadelica* as 'by any standards a treasure house'.[18] Carmichael's work stands in relation to the modernist pioneers of a new era of Gaelic poetry such as Sorley MacLean, in very much the way that Geddes's *Scots Renascence* essay stands in relation to Hugh MacDiarmid's twentieth-century Scottish literary renaissance.[19] Indeed, *Carmina Gadelica* can be seen as an act of linguistic conservation and renewal which is comparable in its own domain with Geddes's interventions in the Old Town of Edinburgh. In both cases much was preserved that otherwise would have been lost and a new cultural dynamic was made possible. Ironically, bearing in mind my comments about the linguistically destructive aspect of William Sharp's insistently Anglophone Celticism, Carmichael's English text has assumed an importance in its own right. His achievement in English can be seen alongside Arthur Waley's translations from the Chinese and Rabindranath Tagore's translation of his own Bengali. As with the publications of Patrick Geddes and Colleagues, the quality of *Carmina Gadelica* extends to its design and production. In common with *The Evergreen*, its printing had been entrusted to T. and A. Constable; indeed, Carmichael makes specific mention of Walter Blaikie's involvement.[20] One can note that Carmichael's *Evergreen* article of 1895 had an initial letter designed by Helen Hay – probably one of the designs on show in the 1896 Old Edinburgh School of Art exhibition – and I would suggest that design provided the impulse for one of the visual glories of the first edition of *Carmina Gadelica*, the intricate designs for the initial letters. The initials were designed by Carmichael's wife, Mary Frances Macbean, and they drew their inspiration both, as in the case of Helen Hay's work, from the time of *The Book of Kells* and from later Highland and Irish work up to the sixteenth century. In his introduction, Carmichael notes that sources of these Celtic letters were manuscripts 'chiefly in the Advocates library'. He goes on to note the extreme difficulty of the task of copying because of the defaced condition of the originals.[21] The Advocates Library provided the foundation for the collection of the National Library of Scotland, and inspection of the occasional decorated initials to be found in that holding of Gaelic manuscripts leads one to conclude that Mary Carmichael was no mere copyist. Clearly she was an artist of ability who transformed limited manuscript materials into a magnificent set of Celtic revival initial letters.[22] Other decorations by her in *Carmina Gadelica* stem from the engraved designs on Pictish sculpted stones of the eighth century reproduced in volume two of John Stuart's *Sculptured Stones of Scotland*. This was published in 1867 by the Spalding Club, which also published a facsimile of another of Mary's visual sources, *The Book of Deer*.[23]

One can further underline the links between the circles of Carmichael and Geddes by pointing out that when Carmichael published his dual-language text of two versions of the Deirdre[24] legend in 1905, both orally collected in Barra in March 1867, the frontispiece was not only by John Duncan, it was the gift of the artist.[25] Duncan gives another indication of his closeness to the Carmichael family in a letter to Geddes, written some years later. He notes that the original of his drawing *Columba on the Hill of Angels*, which was used by Geddes in 1912 as the frontispiece for Victor Branford's *St Columba: a Study of Social Inheritance and Spiritual Development*, is in the possession of Ella Carmichael, daughter of Alexander Carmichael and wife of the then professor of Celtic at Edinburgh, W. J. Watson.[26] Duncan's *Columba on the Hill of Angels* was first reproduced in 1904 in *The Columba Scrip*, a book issued to support the new Church of St Columba in Blackhall, Edinburgh. Duncan's dedication below the image is 'To my Celtic Muse in homage', and it seems likely that this is a dedication to Ella Carmichael for Ella had been the model for the 'muse' figure in Duncan's *Anima Celtica* in 1895.[27] She was one of the most important Gaelic scholars of her generation and had a fundamental role in making sure that her father's *Carmina Gadelica* was published. She was also an informant (as was her father) for Edward Dwelly's invaluable Gaelic-English dictionary, published in parts from 1902 to 1911. Dwelly makes clear the Celtic revival context of his dictionary with a title page complete with Celtic interlace and an illuminated initial similar to those made for *Carmina Gadelica*.[28]

Another link between the Carmichael and Geddes circles is to be found in the person of the musician Marjory Kennedy-Fraser, who had been involved in Geddes's summer meetings in the late 1880s and had in due course taken over organisation of the musical side of those events. That association was seminal in developing her interest in Gaelic song. On the closing page of her autobiography she acknowledges 'Alexander Carmichael, who blazed the folk-lore trail in the Outer Isles; John Duncan, the Celtic painter, who found for me the spot on which I could best begin work; and lastly Patrick Geddes, the pan-instigator, so to speak, who prodded us all into setting forth on the "Hill Difficulty" '.[29] That last reference is to John Bunyan's *Pilgrim's Progress* and it thus resonates with the comments already made about Geddes as 'the interpreter'. It also reminds one that Marjory Kennedy-Fraser was the granddaughter of David Kennedy, a noted precentor in the United Presbyterian Church. Her father, another David, was an internationally renowned singer of Scots song. Marjory Kennedy-Fraser describes her father's ambition as 'to sing the songs of Scotland to the exiled Scots throughout the world'.[30] His notion of how this ambition should be approached was by adapting the songs of Scotland for the drawing-room and the concert platform, and it was this model that Marjory was to adopt for Gaelic song. She has been criticised for this; for example, Sorley MacLean wrote in the 1930s of 'Mrs Kennedy-Fraser's travesties of Gaelic songs'.[31] The problem was that Marjory Kennedy-Fraser's versions were becoming accepted as the authentic music of the Scottish Gael. But to describe them as travesties is only fair comment in the same sense that it is fair comment to claim that Beethoven's setting of Robert Burns's *The Lovely Lass of Inverness* is a travesty, that is to say it is a travesty if and only if it is held to be the original. Not least as a result of Sorley MacLean's own work, standpoints have moved on and the songs can now be heard in their original form as well as in Kennedy-Fraser's transformations. Her Celtic revival work can be seen as part of a movement of European folk-related classical art song rather than as an attempt at authenticity.[32] The appreciation of her work in Geddes's milieu is clear from a collection of tributes to her published after her death in 1930.[33]

Unlike Alexander Carmichael, Marjory Kennedy-Fraser was not a native Gaelic speaker, the last native speaker in her family being her maternal grandfather. But she used some of Carmichael's words and from 1908 worked in close collaboration with Carmichael's follower Kenneth Macleod.[34] Kennedy-Fraser had performed Gaelic song as early as 1882, enthusiastically supported by both her father and John Stuart Blackie.[35] However, in her own words, 'it was not until 1895, for the Summer Meeting Celtic recitals, that I tried it [i.e. Gaelic song] with piano accompaniment, somewhat on the model of the Art Song'.[36] That musical dimension of Geddes's circle is underlined by a publication from 1897, a joint Scottish-Irish publication by Patrick Geddes and Colleagues.[37] The work in question, *Deirdre*, is by one of the key literary figures of the Irish Celtic revival, T. W. Rolleston, who has already been noted for his contribution to *Lyra Celtica*. This is the text of the 'Feis Ceoil Prize Cantata', which had music (not reproduced) by the Irish composer Michele Esposito. It is immediately recognisable as an *Evergreen*-related publication; indeed, its cover design uses Duncan's decoration for the title page of *The Book of Summer*. It shares *Evergreen* format and typography, and reuses initial letters, head pieces and tailpieces by Helen Hay, Nellie Baxter and John Duncan. It also contains work not seen in *The Evergreen*, including an illustration by Althea Gyles.

Published that same year was *The Evergreen Almanac*, a work that concludes three extraordinary years of Celtic revival publishing by Patrick Geddes and Colleagues. The *Almanac* is a primarily visual summary of the ethos of *The Evergreen*. Inside the front cover is the image of *Lapis Philosophorum*, uniting art and science, and inside the back cover is *Arbor Saeculorum*, complementing temporal and spiritual aspects of history. Between those two generalist manifestos are included all the *Evergreen* almanacs – three by Helen Hay, one by Nellie Baxter – and a selection of other work by John Duncan, James Cadenhead and others. Although the emphasis is on images, the accompanying text contains a useful synopsis of the *Evergreen* project, reproducing the defining 'Prefatory Note' from *The Book of Autumn* which outlines the localist internationalism of Geddes's Celtic revival and the reflective summary of the project first printed as the 'Envoi' at the end of *The Book of Winter*. An important addition is a page of text printed on the inner surface of the back cover. It both describes the Edinburgh Summer Meeting and is one of the earliest printed accounts of the Outlook Tower. Its inclusion emphasises the fact that Geddes's Celtic revival work was at the heart of his wider educational and publishing projects. Those interdisciplinary summer meetings are considered in the next chapter in relation to that most potent symbol of Geddes's Scottish generalist ethos, the Outlook Tower.

8

Interdisciplinarity and Cultural Revival

AN INTERNATIONAL SUMMER MEETING

BY THE MID-1890S THE annual summer meeting at University Hall was defined by its interdisciplinary character. Geddes had first experimented with a summer meeting devoted to marine biology in 1886.[1] As noted in chapter five, Fridtjof Nansen had been one of the students. In due course the meetings developed from that starting point to unite numerous disciplines from local, national and international points of view; the international emphasis was reflected in the range of those who attended and taught. A thirty-eight-page prospectus for the month-long summer meeting of 1896 gives insight. The cover makes the interdisciplinary point: 'A Summer School of Art and Science. Social Science, Philosophy, History, Psychology, Education, Civics, Biology, Geography, Fine Art, Music'.[2] The cover also has the University Hall motto *Vivendo discimus* and Geddes's three-dove symbol (of which more below).[3] The year before, in August 1895, Elizabeth and William Sharp had been in Edinburgh so that William could teach a course on 'Art and Life' at the international summer meeting of that year.[4] Elizabeth recalled that 'students came from England, Scotland, France, Italy and Germany; among the lecturers in addition to Professors Geddes and Arthur Thomson were Élisée Reclus the geographer and his brother Elie Reclus, Edmond Demolins and Abbé Klein'.[5] Central to the summer meetings was the interplay of different areas of knowledge. The prospectus for August 1896 advertises Geddes himself teaching courses on 'Contemporary Social Evolution' and 'Scotland: Historical and Actual'. Others teaching included Helen Hay, giving a course on 'Celtic ornament and design', and Élisée Reclus on 'The evolution of rivers and river civilizations'. Music was in the charge of Marjory Kennedy-Fraser, at that time beginning her experiments with setting Gaelic song.[6] The internationalism of the meeting is underlined by the fact that Reclus's course was advertised in French. Adding to the mix from a German perspective was Professor Rein lecturing on 'Herbart and his Educational Theories'. The prospectus notes helpfully that 'many students find difficulty in making notes of the lectures they hear' and goes on to address that by stating that during the summer meeting, 'a little journal will appear every day, under the title of "The Interpreter", which will contain abstracts of the lectures delivered on the previous day'.[7] It gives detailed insight into the content and ethos of the meetings, not least through a weekly reflective column on the preceding week's lectures, drawing links and analogies between them. One of these weekly columns gives an intriguing glimpse. The writer addresses Helen Hay, asking her if she can find in her Celtic ornament 'means for the pictorial representation and symbolism of current ideas'.[8] That challenge must be seen both in the context of Hay's work for *The Evergreen*, and with respect to Geddes's comments (see chapter six) about the Celtic interlace borders undertaken by Hay and her colleagues in the common room at Ramsay Lodge.[9] *The Book of Summer*, the third part of *The Evergreen*, had just been published and, as was the case with the books of spring and autumn, published the previous year, it begins with a visual almanac by Helen Hay. These almanacs – already exhibited at the Old Edinburgh Art School exhibition earlier in the year – are remarkable evocations of the appropriate season; indeed, they give *The Evergreen* a visual identity to complement its description as 'a northern seasonal'. Thus they already address the request for the pictorial representation of ideas. However, the specific reference to Celtic ornament shows that Geddes was concerned to explore further the possibilities of Hay's work. It may be that it was around this time that Geddes was experimenting with adapting the form of the Celtic cross to the purposes of what later became known as his notation or chart of life.[10] Hay was at the forefront of developments in graphic art. Her work invites comparison with the contemporary work of Margaret Macdonald and Charles Rennie Mackintosh in Glasgow.[11]

Further insight into the summer meetings can be gained by considering the members of the consultative committee listed in the prospectus for 1896. Among them was Simon Somerville Laurie, professor of education at the University of Edinburgh, whose relevance to Geddes has been noted in chapter one. Laurie's commitment to Comenius's thinking is significant here, for Comenius was an advocate of visual methods; indeed, in his *Orbis Pictus* he developed for the modern era the notion of visual experience as integral to verbal explanation.[12] For Laurie, in *Orbis*

Pictus 'Comenius applies his principles more fully than in any other, for we have not only a simple treatment of things in general, but of things that appeal to the senses, and along with the lessons we have pictures of the objects that form the subjects of those lessons'.[13] Although Geddes was of a younger generation, he and Laurie, along with David Masson, were among the first lay members elected to the Scottish Arts Club in 1894.[14] It seems likely that through his close connections with Laurie, Geddes had an awareness of Comenius long before he wrote about him in *Cities in Evolution*, and certainly the way he developed his Outlook Tower can be thought of as a kind of three-dimensional *Orbis Pictus* in so far as it was, to echo Laurie's description of Comenius's work, not only a treatment of things in general, but of things that appeal to the senses. The tower was a physical expression of the interdisciplinary philosophy of the summer meeting: 'a centre of research and post-graduate studies in Geography, History, and Social Science, for instruction in Nature-Study, and . . . the beginning of an Index Museum; also for the accommodation of the Edinburgh School of Sociology'.[15] [Figure 5.]

Another of Geddes's consultative committee for the 1896 summer meeting was the marine biologist Sir John Murray. Famous for his work on the *Challenger* expedition, Murray was a key presence in the Scottish Geographical Society and founder of the marine laboratory at Granton, which had hosted Geddes's very first summer meeting. Advising on economics was the Comtean J. K. Ingram of Trinity College Dublin, who had given Geddes early teaching experience in 1881. A number of Geddes's Dundee colleagues and friends were also advisors, among them Martin White, Frank Young and R. C. Buist, but also J. E. A. Steggall, professor of mathematics at University College Dundee. The linen manufacturer Henry Beveridge was another member of the committee; John Duncan and Charles Mackie would soon be at work on murals for his home at Pitreavie Castle near Dunfermline.[16]

The Evergreen Almanac contains a concise description of the summer meeting, which is worth quoting at length:

> The educational aims of the Meeting are best indicated by its actual development. The courses of Seaside Zoology and Garden Botany with which it began in 1886 were complemented the year following by one on the Theory of Evolution. The next step was the introduction of the Social Sciences. Here a general course was arranged, leading from the social phenomena observable in Edinburgh to their abstract aspects, historic, economic and ethical. The parallel teaching of the natural and the social sciences, thus begun, has steadily developed in subsequent years, and always on the same method of reaching general ideas as far as possible through personal observation,

the former deepening as the latter widens. The association of science with practice, of thought with action has similarly been kept in view. Hence the local natural history excursions to systematic 'Regional Survey', artistic and topographic, geological and botanical, then also agricultural and economic, &c., has been a natural progress.[17]

Geddes then emphasises the importance of place and culture: 'from Old Edinburgh and its memories and associations, to the larger life of Scotland, historic or incipient, from Scotland to England and Empire, or from Scotland again to France, and Europe generally' and on to 'the general History of Civilisation'. Each having 'its associated special courses, Anthropology, Education, and Psychology, &c.', which are 'natural steps towards more intelligent interest in that Contemporary Social Evolution in which each is at once actor and spectator'.[18] Finally, both the internationalism and the interdisciplinarity of the meeting are underlined by the varied origins and viewpoints of the lecturers:

> Thanks to the cordial co-operation of many lecturers, belonging to different countries and to the most contrasted schools of thought, this attempt at harmony and unification in studies has made steady progress; and the essential conception, that of experimentally working out, in the freedom of vacation, plans and methods, adaptable to wider application and to more regular educational use – has been justified by the experience alike of students and lecturers.[19]

Philip Boardman characterised the intellectual quality of such events when he wrote of Geddes holding 'constantly before both teachers and students the single goal of reuniting the separate studies of art, of literature, and of science into a related cultural whole' and noting that Geddes's goal was to 'serve as an example to the universities still mainly engaged in breaking knowledge up into particles unconnected with each other or with life'.[20]

THE OUTLOOK TOWER AS GENERALIST COLLEGE

While the annual summer meeting was the defining pedagogical activity of University Hall, the key location for energising that process was the Outlook Tower. It was both at the heart of the social spaces of the residences and central to the wider historical and geographical context of the city. Geddes first began to experiment with the tower in 1892, and as early as 1899 Charles Zueblin of the University of Chicago described it as the world's first sociological laboratory.[21] Geddes's links with the University of Chicago during the period of its foundation have been clarified by the research by Annie V. F. Storr, who has drawn attention to correspondence between Geddes and the President of the newly formed university, William Rainey Harper.[22] It shows that Geddes would have accepted a chair at the university in 1891, had the financial deal enabled him to move more of his activities from Scotland. In the event, Geddes's arguments for more money seem to have been accepted but by that time he had moved on in his activities to the development of the Outlook Tower. It seems that the resources for the chair that Geddes would have occupied in Chicago went instead to the chair of philosophy, whose first occupant was to be John Dewey. But Geddes's close link to that university gives context to Zueblin's comment. In his essay *The Civic Survey of Edinburgh*, published at the Outlook Tower in 1911, Geddes amplified the idea of the tower as sociological laboratory when he wrote of 'learning to view history not as mere archaeology, not as mere annals, but as the study of social filiation', emphasising 'the determination of the present by the past; and the tracing of this process in the phases of transformation, progressive or degenerative, which our city has exhibited throughout the various periods of its history'.[23]

In the context of the summer meetings, *The Evergreen Almanac* refers to the acquisition and development of the tower and its 'adaptation to the requirements of the Summer Meeting, partly as a small Type-Museum, and partly as a centre of teaching, reference and study, in Geography and History, in Social and Moral Sciences, Education, &c.'.[24] This passage leads on to the earliest comprehensive description of the thinking inherent to the arrangement of the Outlook Tower. Again, it is worth quoting at length for its illumination of Geddes's practical interdisciplinarity:

> While current education is mainly addressed to the ear (whether directly in saying and hearing, or indirectly in reading and writing), the appeal of this literal 'Outlook Tower', or Interpreter's House, is primarily to the eye, and the attempt to link the various departments of study, of the Summer Meeting, &c., is thus becoming more and more a matter for actual observation in picture and map, in diagram and 'thinking machine'.[25]

Geddes used the term 'thinking machine' frequently. In its simplest form such a 'machine' was one of Geddes's three-by-three matrices based on the interacting categories of place, work and folk, which in due course led to the complexity of his notation of life. But the crucial thing is that however simple or complex, these were not simply static classification systems, they were *machines*; that is to say they depended on movement. Thus for Geddes the use of a thinking machine implied the development and evolution of understanding in an interdisciplinary framework. And furthermore that framework – whether a matrix of cells or the Outlook Tower itself – took account of both the theoretical and the practical. By living we learn indeed, *Vivendo discimus*, in the widest and most profound sense. Geddes thus used the word 'machine' to imply a method for controlling the movement of thought that had the potential to create new understandings. He continues:

> The utilisation of the Tower thus best begins in its Camera Turret, with the actual landscape. From the Art of Seeing with the help of open eyes and the Camera Obscura, one proceeds naturally to the Seeing of Art by the help of representative pictures, so that, in some measure, every one may become his own Art-Critic, appreciating contemporary evolution in art with freshened eyes. Parallel to this comes the Art of Seeing in Science – Observation and Reasoning upon the phenomena themselves, independently of book or other hearsay accounts of them.[26]

The account then shifts discipline, returning again to 'the actual Prospect, but now from the geographer's point of view, and with his model and map now replacing camera and picture'. Then the usefulness of specialisation is noted, but only as a necessary expedient within a wider sociological vision of the evolution of societies:

> Further comes the analytic study of this mass of impressions by the help of the different specialist sciences; astronomer, meteorologist, physicist, geologist, botanist, anthropologist, architect, historian, economist, moralist, being treated as successive specialist observers, each elucidating a class of phenomena which he artificially isolates from the rest for the temporary convenience of study. These all return to unity in the study of Contemporary Evolution.[27]

And in that context of contemporary reality Geddes continues, 'thence we descend the Tower through its successive storeys of City, Scotland, Empire (or rather Language), Europe, and World, concrete personal observation thus widening with the results of travel, and deepening with the corresponding thought'. And finally:

> In such ways, then, it is hoped to express the unity underlying all the various Fields of Study – Art and Geography, Biology and Biography, Education and Civics, History and Philosophy, and to express the Unity which underlies these many different products and processes of Contemporary Social Evolution.[28]

The Outlook Tower's location was for Geddes a key to its educational potential. It stands on a ridge of rock created by the combined forces of volcanic activity and glaciation. That ridge underpins the entire

Old Town of Edinburgh, and extends towards the extinct volcano of Arthur's Seat. On one side of the tower is a symbol of temporal power, the castle, on the other a symbol of spiritual power, the Cathedral of St Giles. Even closer is the then symbol of rejection of the authority of the state in matters spiritual, the Free Church College, now New College of the University of Edinburgh. Within easy view was a symbol of national economic power, the headquarters of the Bank of Scotland. And in the valley beneath the tower lies a public garden bisected by a railway that tunnels beneath two major institutes of art, the Royal Scottish Academy and the National Gallery. Across that valley is Princes Street, which, despite its name, in the memorials which line its southern side commemorates not royalty but figures of literary and civic virtue, among them Sir Walter Scott, Allan Ramsay and Thomas Guthrie. Beyond this street is the Enlightenment grid of the New Town, and across that urban area to the north-west are the Highlands, while to the north-east are the ports of Newhaven and Leith and beyond them the firth of Forth. Across that estuary are the hills of Fife.[29] A few miles to the south of the Outlook Tower are the Pentland Hills. Far off to the east lie Bass Rock and Berwick Law. This list is not the enumeration of features from a map, but simply a record of features that it is hard to miss from the Outlook Tower on a reasonably clear day. The tower is thus located at a key point from which to think about the relationship between city, region and wider cultural contexts. Geddes made the whole parapet into a geographical indicator, but whereas most such indicators name only that which can be seen, one finds Geddes taking a regional and national view, for most of his indications are to aspects of Scotland which cannot in fact be seen from the tower. So the observer takes note of the bearings of Iona, Cape Wrath, Stornoway, Oban and the almost visible Stirling, and their mind begins to encounter the regional surveyor's task: not just to see, but to see a geographical and cultural context and to begin to understand cultural evolution and future possibilities.

Complementing the directly perceptual aspect of the Outlook Tower, Geddes considered it to be an index museum, that is to say a museum that was able to potentially classify all knowledges. Key to that notion was that it was also a graphic encyclopaedia and an encyclopaedia of methods.[30] In a letter written in 1905, Geddes explains it in the following terms:

> the Tower may be best explained as simply the latest development of our Edinburgh tradition of Encyclopaedias, and hence arising in turn in the very same street where are all the others, *Britannica*, *Chambers*, and minor ones.

It is in fact the *Encyclopaedia Graphica*. The *Encyclopaedia Graphica* for each science and art in turn and in order'

and he concludes: 'but this makes it next *Encyclopaedia Synthetica*'.[31]

As one would expect from an *Encyclopaedia Graphica*, visual art was fundamental. This included a frieze painted by James Cadenhead and Helen Hay, and the use of stained-glass windows for teaching purposes. The designs of two of these, the *Lapis Philosophorum* and the *Arbor Saeculorum*, date from 1894 and 1895, and have already been noted with reference to their appearance in *The Evergreen*. There were two others, one a valley section given the title of *The Typical Region: Mountain to Sea, with the Fundamental Occupations*. [Figure 6.] The other was *The Good Shepherd*.

According to the 1906 guide to the tower, the *Arbor Saeculorum* window was positioned separately, while the other three windows formed a unified group with the horizontal valley section panel positioned above a doorway and the other two panels forming the left- and right-hand panels of the door itself.[32] There is a drawing by Geddes which shows this arrangement.[33] On it Geddes has written the German word 'stammbaum', which can refer both to a family tree and to a phylogenetic tree, and no doubt both meanings were in his mind.[34] This not only sheds light on the *Arbor Saeculorum* design, but it resonates with the content of a letter from Geddes to Charles Mackie referring to the first *Evergreen* cover. Geddes writes of how pleased he is that Mackie has selected a particular species, *Aloe plicatilis*, as the basis for his design. Geddes notes that as biologists he and Arthur Thomson have a particular evolutionary interest in the plant, and thus he takes Mackie's selection of it 'as an omen that science and art are going to be better friends than ever'.[35] It can be noted that aloes in general are an appropriate symbol of Geddes's commitment to revival, for they have both healing properties and can survive long periods of neglect and drought. Similar in form although different in species is the American Aloe, more commonly called the Agave. The Agave seems also to have a symbolic role here for Geddes; indeed, the difference between the *Evergreen* covers and the *Arbor Saeculorum* can be seen in terms of the difference between the Aloe and the Agave. Both are plants of survival and revival, but the Agave is particularly characterised by a large bud that appears at the top of the plant prior to its infrequent flowering. That would seem to be the inspiration for the bud which tops the *Arbor Saeculorum* image. Geddes would have been aware that *Agave Americana* is also known, due to its infrequent flowering, as 'the century plant', and the resonance of that latter name with the *Arbor Saeculorum* – which should be translated as the tree of the generations or eras – is clear. The 1906 guide details the *Arbor* window in its place in the Outlook Tower as follows:

> Still further down the stair is another stained-glass window. A great tree and its branches, spreading right and left, suggests the twofold aspects of each historic era, temporal on one side, spiritual on the other. The tree has its roots amid the fires of life, and is perpetually renewed from them; but the spirals of smoke which curl among its branches blind the thinkers and workers of each successive age to the thought and work of their precursors. While the branches symbolise the past and passing developments of society, the bud at the tree-top suggests the hope of the opening future. Two sphinxes guard the tree and gaze upward in eternal questioning; their lion-bodies recalling man's origin in the animal world, their human faces the ascent of man. Issuing from the smoke-wreaths at the top of the tree are the phoenix of the ever-renewed body, and the butterfly (Psyche) of the deathless soul of humanity. On either side of the window rises a series of symbols, those on the right hand indicating the dominating spiritual forces of the great historic periods, those on the left the corresponding temporal powers.[36]

In the *Arbor Saeculorum* the temporal and spiritual contexts of the Western tradition are thus presented in a historicist manner from ancient Egypt to the late nineteenth century. A further page is devoted to describing these symbols. For example, 'on the right, the bird of Pallas symbolises Greek wisdom and Greek ideals; on the left the flying galley recalls the essentially maritime character of the Greek empire and civilisation'.[37] The symbols culminate as 'above the flag of Liberty, we find the red flag of Socialism, the black flag of Anarchism, symbolising contrasted tendencies which we have in our midst even now; and corresponding to these, the symbol of the enormous increase of wealth, yet also of empty hands'.[38] Thus we find ourselves back in the anarchist and Marxist company of Mavor, Kropotkin and Thomas Kirkup in Geddes's James Court flat in 1886. The distinction between the two aspects of the *Arbor* is primarily between the Word (whether that of Jehova, Athene, Kropotkin or Marx) and the State. 'Last of all, on either side, we find a query – the eternal question of the Sphinxes for ever reiterated, for ever unanswered; yet above these an opening bud, a flower as yet undefined.'[39] One can recall here that while the *Arbor Saeculorum* reflects on the evolution of cultural histories, the *Lapis Philosophorum* (as discussed in chapter five) encodes the methods of the arts and sciences.

Complementing them was the visual discourse on the interaction of geography and society, the valley section window. Its personal inspiration lay, as noted in chapter two, with Geddes's childhood experience of hill and river, while its intellectual inspiration lay with the work of Élisée Reclus and Frédéric Le Play. Geddes, as ever, synthesised, generalised and made accessible. Here is his description of the valley section's wider significance from *A First Visit to the Outlook Tower*:

> a stretch of landscape from sea to hill-top, passing through wheatlands and oatlands, through pasture and forest, and descending on the other side with a ruder slope. This may be viewed in various ways, for we may think of it as the landscape we have just looked at from the gallery of the Camera Tower, stretching from the blue waters of the Forth to the bare Pentland hill-tops; or we may regard it as a diagram of the great plain of Europe rising slowly to the mountains, or of North America from the Atlantic sea-board to the Rockies and beyond. So that we come to see that the little landscape is a typical section of the Earth at many points, and indeed, as the legend written under it tells us, a 'microcosm of nature'.[40]

This stained-glass window is just one of many expressions of Geddes's valley section, including classification of rural occupations and related urban trades, and even adaptations of the valley section image for classification of types of landscape painting.[41] For Geddes, at some point all our activities can be argued back to those of the valley; even if we live in a highly urbanised, automated society, our lives are still linked back to the activities of miner, forester, hunter, shepherd, farmer and fisher.[42] For Geddes it was important we realise this and indeed participate in those activities if we can, so that we can understand them better. In a letter to Jane Whyte written in October 1898 on the subject of the education of her son, Aird, Geddes outlines his view of the importance of such participation:

> any 'professing' I may be justified in attempting, I would distinctly base not primarily on academic but on solid ground, i.e. on a pretty long silent experience of going up and down hill, literal gardening, literal fishing, literal tree-planting, literal building and the like: seeking, in short, the Wisdom of the Crafts.[43]

On the previous page of the letter is an elegant representation of the valley section, which also includes a direct application of Frédéric Le Play's *lieu, travaille, famille* to the diagram.[44] From today's perspective one might argue that the 'typical region' is out of date as – for example – at present we get more textiles from petrochemicals rather than – say – sheep farming. But one could respond by pointing out that the fact that we are now mining some of the raw material of clothing rather than growing it, simply draws attention to the importance of the basic interactions with the environment that Geddes identified. That is to say that the principle of the valley section is intact, even if the detail is open to change. Thus Geddes provides a model of types of economic interaction between human being and environment: the taking of minerals by the miner, the husbanding of resources by forester and shepherd, the annual cultivation cycle worked with by crofter and gardener and farmer, the wild harvesting of the fisher. And with each activity came its own lore, legends and myth, part of 'the Wisdom of the Craft'.

On another level the stained-glass version of the valley section is a multiple representation of what the physical and social world is at the moment and could be in the future. Looking at the Latin wording which appears below the window – 'Microcosmos Naturae. Sedes Hominum. Theatrum Historiae. Eutopia Futuris' – one sees Geddes insisting on a set of mutually illuminating views of the valley. The valley is first a 'microcosm of nature', but also the 'sedes hominum', the seat of man, the place where human beings make their lives, and linked to that it is the dramatic 'theatrum historiae', the theatre of history. Finally, it is the 'eutopia' or 'good place' of the future. Not 'utopia' with a 'u' by which Sir Thomas More implied an idealised 'no place' but 'eutopia' with an 'eu', a good place that Geddes believed could be achieved through local and international co-operation.[45] But Geddes doesn't stop there. Looking at the window we note that the microcosm of nature – with its suggestions of a more primitive way of life – is linked to the flight of two game birds, while the theatre of history – with its implication of our present, historically conscious, view of ourselves – is represented by fighting birds of prey. But the good place of the future, the 'eutopia', is symbolised by three doves flying in harmony. For Geddes this was symbolic of a transition from a crudely mechanised capitalist society based on unfettered competition (what he called the paleotechnic world) to his concept of a society which made use of sophisticated, environmentally friendly technologies based on local co-operation (what he called the neotechnic world).

Those three doves demand further attention, for they run throughout Geddes's work. Their first appearance is in a stained-glass window for Riddle's Court in 1889.[46] By 1895 they were a key motif for *The Evergreen*, and from that time onwards they were to be found on information sheets, booklet covers and so on, and on the flyleaves of other publications such as those of the Celtic Library. [Figure 10.] For Geddes they were not just symbolic birds of peace, they also represented three concepts beginning with the letter 's' – sympathy, synthesis and synergy. Thus this message is also part of the valley section window in the Outlook Tower. What Geddes meant by 'sympathy' was the

Enlightenment use of this word, interchangeable with 'empathy' as we use it today, that is to say the ability to accurately imagine oneself in another's place. 'Smith wi' his sympathic feeling', as Robert Burns put it, referring to Adam Smith's use of the word in his *Theory of Moral Sentiments*.[47] And from that sympathy, that emotional engagement with other people, Geddes proceeds to the intellectual power of comparing and synthesising different ideas. And finally, having emphasised first the emotional, and secondly the intellectual, he moves to the social notion of 'synergy', that is to say of working together to solve problems and create opportunities – in other words, Geddes's primary principle of successful human society, co-operation.[48]

Thus in that one image from the Outlook Tower, the stained-glass version of the valley section diagram, Geddes provides a visual basis for human ecology not only by drawing attention to the necessity of exploring any environment in terms of its natural characteristics, and the way in which those characteristics relate to the people who live there and the work done, but also by specifying the psychological attitude necessary for any such analysis, that is to say an emotional, intellectual and cooperative engagement with that place, those people and their work. Had Geddes's essentially simple but comprehensive and well analysed ideas been borne in mind more often over the last century, many environmental disasters would have been avoided. Geddes provides us not only with both a set of analytical categories derived from Le Play – folk, work and place – but with a set of emotional, intellectual and social methods – sympathy, synthesis and synergy – for how we should give practical meaning to those analytical categories. So Geddes's three doves should not be seen as merely some kind of logo. They are in fact a manifesto for thinking and practical engagement.

The final window, *The Good Shepherd*, underlines one of the occupations referred to in the valley section. But its implications extend much further. At its foot is the phrase 'flos herbae', which translates into Latin the title of Geddes's *Evergreen* essay *Flower of the Grass*.[49] In that essay Geddes emphasises that a way of life, in this case the pastoral, leads to a particular social formation, the patriarchal, and he contends this is one of the underpinnings of European society. An exploration of such ideas can be found in Victor Branford's *St Columba: A Study of Social Inheritance and Spiritual Development* published in 1912. The biblical reference of the words 'flos herbae' would have been clear to most of Geddes's milieu. It extends the discussion further, for the passages in the Bible which refer to 'the flower of the grass' use this 'flower' as a metaphor for the transitory nature of human life. The most concise expression is in The First Epistle of Peter: 'All flesh is as grass, and all the glory of man is as the flower of grass. The grass withereth, and the flower thereof falleth away: But the word of the Lord endureth for ever.'[50] This revisits Isaiah chapter 40, and the message of death and continuity is the same. For Geddes, a combination of botanical reference and biblical authority made absolute sense. It was not simply a use of biblical reference in aid of pedagogy, but use of the Bible (regardless of one's belief or otherwise) as a profound source of insight. The window also refers to the spiritual-ecological imagery of the first verse of Psalm 23, specified in the guide to the Outlook Tower by reference to 'quiet waters', a quotation from the Scottish metrical version of the psalm.[51]

Such eco-theology is underpinned by Geddes in his essay: 'the theologian who would understand, who would use the Lamb as a sacred symbol, should first feel (ay, and use as teacher) the thrill of its gentle influence as a living thing'.[52] It is also interesting here to find Geddes qualifying his intellectual allegiance to Comte in favour of Le Play:

> Auguste Comte is popularly supposed to be a radical, a democratic man of modern science. But he makes his contributions to sociology from the standpoint of hierarchy of feeling and genius, of the aristocracy of action and thought. Conversely, it is Frédéric Le Play, whose point of view it is that has been followed and developed above, and who is popularly supposed even in his own country to make his appeal to capitalist and conservative, to aristocrat and priest, who has really established for us the vital doctrine of all democracy . . . Comte sees the great stream of Humanity; but in this he calls attention mainly to the Calendar of Great Men, to the men of genius as Her chief servants – for him, proletarian and woman are little better than grown children, to be guided and governed for ever by patrician and priest. But for Le Play, worker and woman unite to form the elementary human family, and from them, not only by bodily descent, but social descent, from their everyday life and labour, there develops the whole fabric of institutions and ideas, temporal and spiritual.[53]

Geddes's holistic cultural and ecological vision was given impetus and focus by the development of the Outlook Tower. A final element was the inclusion in the entrance hall of a globe made under the supervision of Élisée Reclus, and the on-site work of his nephew Paul.[54] It was large enough for the student and visitor to at least begin to experience by eye and hand the actual physical characteristics of the planet. That Geddes/ Reclus insistence on visualising the planet as a sphere in space only really entered popular consciousness in 1968 with the remarkable photograph of the Earth taken from above the surface of the moon by the crew of

Apollo 8. But it was precisely this consciousness of planetary scale and ecological fragility that Geddes was attempting to evoke in the Outlook Tower.[55]

Charles Zueblin's characterisation of the Outlook Tower as the world's first sociological laboratory has been noted; it can be emphasised that Zueblin considered the tower merited that description because it combined features of school, museum, atelier and observatory. Thus it was Geddes's commitment to an interdisciplinary practice that underpinned Zueblin's comment. For the participants at the summer meetings, the Outlook Tower was not just the venue, it was the symbol of educational purpose within the wider living context of University Hall and of Edinburgh itself. And Geddes took it further still. In notes to Victor Branford in 1902, Geddes extended the tower's scope:

> The Tower of Thought and Action needs a corresponding basement and not merely that of Arts and Sciences in general as hitherto but a sub-basement or catacomb proper in which the life of feeling is similarly to be recognised. Is there any escape from this argument? and if not where does it lead? Do not imagine that I have in any way lapsed into a Quietist or delirious mood. This is only a clearer development of what has been more or less incipient all along.[56]

He referred to this later as the 'complemental Inlook' function of the tower.[57] Ever sensitive to the *zeitgeist*, Geddes made these notes two years after the first publication of Freud's *Interpretation of Dreams*, but long before it was translated into English.

The development of the Outlook Tower exemplifies the fact that Geddes was a generalist born into a culture in which generalism was advocated as an educational goal. For Geddes such an approach could be applied on all levels from the individual to the planetary. On an individual cognitive level, a solitary thinker would find that different approaches within their own thinking illuminated one another. On a cultural level, one would have one's thinking illuminated by other thinkers coming from different perspectives. On a societal level, the blindspots of thinking in one nation or region or culture would be illuminated from the standpoints of others. Geddes's Outlook Tower – whether considered as an index of indexes, an *Encyclopaedia Graphica*, the heart of University Hall, a definer of the scope of international summer meetings or the starting point of regional survey – was an architectural statement of such critical generalism, 'a Tower with a view over everything in life' as Elspeth Davie called it.[58] It is interesting to note that in 1912 Geddes made a more conventional suggestion for such a generalist architectural statement, namely a National Library of Scotland. His proposals were close to what in the end came into being, when the National Library was constituted by an Act of Parliament in 1925.[59]

9

Paris 1900: Local Revival as Global Project

THE PARIS UNIVERSAL EXHIBITION

PHILIP BOARDMAN WRITES OF Geddes as an ardent Scot who 'was an equally loyal and intellectual son of France'. He continues, 'Yet with the approach of the twentieth century he placed himself no less convincingly in a third category, of world citizenship, without abandoning either France or Scotland.'[1] The practical side of his internationalism was underlined by the transformation of the Edinburgh summer meeting into the Paris International Assembly, which was Geddes's educational complement to the grand spectacle of the universal exhibition of 1900. France had been an intellectual home from home for Geddes since his student days and his French links had been recognised by academic colleagues from the 1880s onwards. For example, S. S. Laurie advising a student travelling to France had singled out Patrick Geddes 'as the Scot who had best kept up the French connection' in matters academic and cultural.[2] The student was Stewart Robertson and Geddes provided him with numerous academic introductions in France, including to the pioneering French educational theorist Gabriel Compayre, whose *Histoire des Doctrines de l'Education* was the only work of its kind available at the time.[3]

Context can be given to Geddes's educational intervention at the 1900 Paris exhibition by noting the strand of French, Belgian and Breton contributions to *The Evergreen* a few years before. These included images strongly influenced by developments in French art on the cusp of symbolism and post-impressionism. The first such image to be published in *The Evergreen* was *Pastorale Bretonne* by Paul Sérusier, which appeared in the first volume. Sérusier was an inspiration to Charles Mackie, who contributed two illustrations to that volume, along with the cover and title page designs, and it was Mackie's contact with Sérusier that led to his contributing to *The Evergreen*. As noted in chapter six, Claire Willsdon has raised the intriguing possibility that Sérusier, like Mackie, painted a mural for Geddes's flat in Ramsay Garden.[4] Sérusier's image was immediately followed by a piece by Dorothy Herbertson entitled 'Spring in Languedoc' and a little later is a piece in French, 'La Littérature Nouvelle en France', by Charles Sarolea of Edinburgh University. In the second volume of *The Evergreen* the Francophone cultural aspect is also strong. There one finds William Sharp's translation of 'Les Flaireurs' by the Belgian writer Charles van Lerberghe.[5] A few pages later is a piece, 'Le Dilettantisme' by Abbé Félix Klein, published in its original French. It is immediately followed by Edith Wingate Rinder's English version of the Breton legend 'Amel and Penhor'. There also is another piece in French, 'La Cité du Bon Accord', by Élisée Reclus, who is described in a footnote as 'the joint-apostle with Tolstoi of the higher Anarchism'. Again in this volume, Mackie's illustrations show their Pont-Aven influence. Another image, by the Glasgow painter E. A. Hornel, takes as its subject a Japanese woman. The title, however, is *Madame Chrysanthème*, a reference to the story by the French writer Pierre Loti. The third *Evergreen* volume, *The Book of Summer*, was published in 1896 and contains another version of a Breton legend, 'Telen Rumengol', by Edith Wingate Rinder, and another piece in French, 'Vers l'Unité', a meditation on the common aspects of the major global religions, by Abbé Klein. Along with Mackie's images is one by Robert Brough, another Scottish artist with close links to France and, indeed, to Brittany. Two more, which are given French titles, are by Andrew K. Womrath, an American artist who made his reputation in France. Here also one finds John Duncan's Joan of Arc illustration, *The Way to Rheims*. The final volume of *The Evergreen* contains Paul Desjardins's 'Il neige', a further Breton legend by Rinder, 'Saint Efflamm and King Arthur', a piece by Élisée Reclus's brother, Elie Reclus, entitled 'Pourquoi des Guirlandes Vertes à Noel', and 'A Devolution of Terror' by Catherine A. Janvier. Also here is a piece on Breton custom by M. Clothilde Balfour, 'The Black Month'.

Thus *The Evergreen* is a product of the Scottish Celtic revival which has close to its heart a current of French, Breton and Belgian culture and commentary. In all the volumes there are illustrations by Charles Mackie, in which he shows his strong Breton links.[6] More widely, Clare Willsdon has noted the affinities of both Mackie and John Duncan 'with Continental Symbolism's ideology of mural painting as an affective medium affording entry into dreams and emotional states', and she goes on to note Duncan's reading of Sar Péladan's theories, as well as those of Burne-Jones.[7] Duncan's stylistic debt to the French mural artist Puvis de Chavannes is also clear. It is given context by the publication by Patrick Geddes and Colleagues, in about 1896, of versions of four decorative panels from Puvis's frescoes of St Geneviève in the Panthéon at Paris. Such conjunction of art and city development set an example to Geddes; indeed, these images were published at exactly the time Duncan was engaged in his Ramsay Lodge murals.[8] In an *Evergreen*-related painting, *Joan of Arc and her Scots Guard*, completed in the mid-1890s, Duncan and Geddes had drawn attention to the links between France and Scotland from the perspective of history. It was originally located in the common room of the University Hall residence of St Giles House.[9] It was first mooted by Geddes in a letter to Duncan sent in November 1895, in which he notes a suggestion from Andrew Lang 'of the Franco-Scottish society' for a painting of Joan of Arc with her bodyguard of Scottish Archers.[10] As noted in chapter 6, a drawing of the design appeared in *The Evergreen* in 1896 under the title *The Way to Rheims*. It has the caption 'Franco-Scottish Society – Sorbonne, 18th April 1896', the date which marked the first meeting of the new society in Paris.[11] Consideration of Andrew Lang can extend perception of this Francophile milieu further, for as already noted, in 1896 Lang was bringing to fruition another project related to Joan of Arc, the publication of his novel *A Monk of Fife*.[12] It tells the tale of a Scottish cleric from St Andrews in the France of St Joan, and it was illustrated throughout by the Arts and Crafts designer Selwyn Image.

Complementing his sympathies with French culture was Geddes's regard for the social value of major international exhibitions. For him such intelligently structured public spectacles were a starting point for creative thought; indeed, Alessandra Ponte, in her account of the place of the Outlook Tower in Geddes's thinking, has noted that he regarded universal exhibitions as the 'primordial liquid' which gave birth to museums.[13] In 1887 he had published an extended critique entitled *Industrial Exhibitions and Modern Progress*, in which he exhorted the organisers of future industrial exhibitions to 'take real and detailed heed of the claims of Art, Science, and Political Economy'.[14] The Paris universal exhibition of 1900 seems to have met Geddes's prescription to a significant degree. James Mavor was in Paris at the time and he comments that the exhibition illustrated Geddes's theme more fully than any other yet held, quite apart from Geddes's own contribution.[15] It was centred around a series of national pavilions constructed alongside the Seine. Geddes made strenuous but unsuccessful efforts to have this *Rue des Nations* preserved as a feature of Paris after the exhibition, proposing that each pavilion would become an international museum devoted to a different subject. The wider political and cultural context for Geddes's participation in the events associated with the Paris exhibition, including the intricacies of the Dreyfus case and the international response to it, has been explored by Siân Reynolds. She makes the point that Geddes's political and social networks in Paris, not least his radical political links, did not help his aim of preserving the *Rue des Nations* as a symbol of internationalism and peace, for this was a time of growing French right-wing sentiment.[16]

One of the proposed but unexecuted exhibits at the Paris exhibition was a great globe designed by Élisée Reclus. The point of Reclus's globe was that it should be big enough for visitors to be able to recognise the geographical features of the planet itself. Thus his intention as a geographer was to enable viewers to place themselves in relation to the planet. Like the Outlook Tower, the globe 'was conceived as a permanent teaching instrument that could be updated as the most recent data arrived from geographical explorations and surveys'.[17] It was 'a powerful metaphor for the cosmic unity of human life' and it implied connectedness regardless of political boundaries, race or creed. In short it testified to a common humanity.[18] As noted in the previous chapter, in the 1890s Paul Reclus, Élisée's nephew, had worked on a globe project while staying at the Outlook Tower.

Geddes's Paris International Assembly was an ambitious undertaking. In essence the activities and educational ethos of the Outlook Tower were transferred wholesale to France. An extensive lecture programme began in mid-May and continued to the end of October, and lecturers included Outlook Tower regulars such as Geddes himself, Thomas Marr and John Duncan. In addition there were a number of American speakers and this transatlantic involvement reflected a new dimension of Geddes's internationalism, which he had begun to develop in earnest as part of his efforts to gather support, both financial and intellectual, for the International Association, a body which Geddes had created to support the Paris International Assembly.[19] Geddes's American contacts included the psychologists Stanley Hall and William James, and the educator John Dewey. Another was the social reformer and educator Jane Addams, who with co-founder Ellen Gates Starr had in 1889 established Hull House in Chicago along lines suggested by Toynbee Hall in London, a development that had also influenced Geddes. Addams lectured for Geddes in Paris, and the Chicago link was underlined by John Duncan's subsequent appointment as professor of art at the newly formed Chicago Institute, where he worked until 1903. Before leaving Chicago, Duncan was closely involved in Hull House, becoming a Resident at the settlement and painting a mural panel of Tolstoy towards the end of his stay. A memoir of Tolstoy by Jane Addams published in *McClure's Magazine* reproduces Duncan's image.[20] A year after the Paris exhibition, Frank Lloyd Wright would use a speech at Hull House to synthesise an Arts and Crafts sensibility with a respect for the machine.[21] If Duncan was present, Wright's approach to the traditional values of arts and crafts on the one hand and his interest in machine-ethic modernism on the other would certainly have reminded him of Geddes.[22] A letter from Geddes to Jane Addams indicates that Duncan was expected to return to Chicago in late 1903 to complete the Hull House murals. Geddes apologises to Addams for 'forcibly retaining reluctant Duncan' to work with him on what was to become the influential report for the Carnegie trustees in Dunfermline, *City Development*.[23] The letter underlines the close links of both Geddes and Duncan to Addams and her circle. The value of Duncan as part of that milieu is at last becoming clear, not least with respect to Addams's close friend and collaborator Ellen Gates Starr.[24] For Addams, meeting Geddes in Paris in 1900 seems to have been important in the formulation of her peace activism. Geddes's grass-roots approach can be discerned when he contrasts the work of 'The diplomatists, the journalists' who 'have had their Peace Congress at the Hague' unfavourably with the 'more real Peace Congress', the 'gathering of workmen and architects and artists' who 'have silently made peace more interesting than war' through the International Assembly.[25]

Siân Reynolds has noted that the French word 'intellectuel' was coined in the 1890s to refer to 'those who lived by the intellect but who applied it to a wider world than the academy – often taking a public stance on matters of principle'.[26] This seems a most appropriate description of Geddes. James Mavor provides further insight into the wider intellectual milieu of which Geddes was part in Paris. Mavor himself made a number of visits to Rodin in his studio and recalls meeting Élisée Reclus by chance at a lecture on Rodin given by the art critic Camille Mauclair.[27] Mavor's reminiscences of Geddes in Paris are intriguing, not least his account of an excursion around the city along with the assembled Geddes and Kropotkin families.[28] Their guide was an elderly Parisian communist named Le Français who had been behind the barricades as a boy in 1830, and again in 1848 and 1871. Mavor notes that Le Français showed them 'how certain parts of Paris were readily convertible into miniature fortresses'. This has more than just an obvious historical interest, for such information gives real insight into the origins and development of the city, the nature of its communities and how they are physically defined, topics of central interest to Geddes as a pioneer of town planning. Geddes and Mavor took such considerations further in a walking tour in northern France in the late summer of 1900, reflecting on the *ronds-points* of the forest of St Cloud, which afforded rendezvous for hunters, as the putative origin of the plan of the city of Paris.[29]

LOOKING TO INDIA: GEDDES, VIVEKANANDA AND NIVEDITA

Mavor also comments on the presence in Geddes's circle in Paris of Swami Vivekananda and his disciple, the Irishwoman Margaret Noble, the latter better known as Sister Nivedita. Mavor's account is not wholly accurate, for he assumes that Vivekananda and Nivedita first met that year in Paris, whereas in fact they had met some five years earlier in London,[30] but the important point here is that they were closely connected with Geddes in Paris. Seven years earlier, at the World Parliament of Religions in Chicago, Vivekananda had 'presented Hinduism to the world at large as a major religion, emphasising its antiquity'.[31] Taking his lead from the ideas of Ramakrishna, Vivekananda articulated the case for Hindu revival. In Paris he lectured on Indian art, rejecting theories of Greek influence to underline the independent value of the early Buddhist art of India.[32] In due course Nivedita was to develop this position, as was, in a more systematic way, Ananda Coomaraswamy. Vivekananda died in 1902 but Nivedita continued to be one of Geddes's intellectual allies up until her death in 1911. Also present in Paris was the eminent Indian scientist Jagadis Chandra Bose. Sister Nivedita numbered his wife Abala – who was herself a social activist of note – among her closest friends, and twenty years after their Paris meeting Patrick Geddes was to be Bose's biographer. Initial contact between Geddes and Nivedita had been made early in 1900, not in Paris but in America, where Nivedita had been teaching with Vivekananda. The same year saw the publication of Nivedita's first important contribution to the Hindu revival, *Kali the Mother*.[33] In a memoir of Nivedita printed in *The Sociological Review* in 1913, Geddes recalled their first meeting in New York and how it 'continued into intimacy and collaboration during the following summer, at the meeting of the International Association which became the Summer School of the Paris Exhibition of that in many ways memorable year'.[34] Nivedita had much in common with Geddes. Both complemented their dedication to cultural revival with a passionate interest in new educational methods; indeed, prior to her involvement with Vivekananda, Margaret Noble had been a respected advocate of the methods of Pestallozzi and Froebel.[35]

The link between Geddes and Nivedita in the Paris of 1900 thus places in close touch with one another a central figure of the Celtic revival in Scotland and one who was to become a central figure of the Hindu revival in India. Geddes emphasised their common methods when he wrote in a memoir of Nivedita that her career could not be fully appreciated 'without some corresponding grasp of the geographical outlooks and evolutionary methods which she so clearly held'.[36] He then points out that she considered these methods as 'only second in their significance and value to the philosophic and religious synthesis of her adopted order, and as an essential instrument of its social and educational purposes'. He continues that these two perspectives – 'let us call them of Le Play and Ramakrishna' – became for Nivedita 'closely and vitally related, as fundamental and supreme, alike indispensable to the understanding of the social body and the social mind, their nature and their spirit', regarding this as 'one of the main clues to her rare range of sympathy and understanding, at home as she could be either in Paris or in Calcutta, and to her essential life-pilgrimage, from West to East'.[37] Geddes goes on to quote from Nivedita's *The Web of Indian Life*: 'The foundation stone of our knowledge of a people must be an understanding of their region. For social structure depends primarily upon labour and labour is necessarily determined by place. Thus we reach the secret of thought and ideals.'[38] Nivedita's commitment to the Le Play-Geddes principles of place, work and folk, and to Geddes's associated notion of regional survey could hardly be more emphatic. She makes her personal debt to Geddes clear in an epigraph in which she thanks him for 'teaching me to understand a little of Europe' and thus giving her a method by which to read her Indian experiences.[39]

Just as Geddes's work in Scotland must be seen in a pan-Celtic context, so the activities of Nivedita must be seen in a pan-Asian context. Crucial here was Nivedita's link with the Japanese art critic and teacher Kakuzo Okakura, who she met in Bengal in 1902.[40] Okakura himself had no direct link with Geddes but Kiyoshi Okutsu has drawn an illuminating analogy between the thinking of the two men.[41] This analogy is given historical weight by the fact that Okakura's brother, Yoshisaburo Okakura, was supported by Geddes's benefactor Martin White when he gave a course of lectures at the University of London. They were published under the title *The Japanese Spirit* in 1905.[42] Kakuzo Okakura's *Book of Tea* was published the following year in New York.[43] A more direct link between Geddes's thinking and the Japan of this period has been pointed out by Toshihiko Ando. He notes that the first intensive reading of Geddes by a Japanese thinker of public note was that of the novelist Natsume Soseki.[44] This is an intriguing conjunction for it places Geddes's ideas at the very heart of the development of modern Japanese literature. Kiyoshi Okutsu has furthered Ando's insight by exploring Soseki's comments on Geddes's book (co-authored with J. Arthur Thomson) *The Evolution of Sex*.[45] This reading of Geddes is given further interest when one notes that while Soseki was living in London from 1900 to 1902, he travelled to Scotland at least twice. Damian Flanagan has noted that 'Scotland was an important place to Soseki . . . He was largely miserable in London, but wrote glowingly of Scotland.'[46] In 1901 he signed a postcard, postmarked Edinburgh, showing the birthplace of Robert Burns.[47] This opens up the intriguing possibility not only that Soseki visited Burns's birthplace in Ayrshire, but that he visited the Outlook Tower in Edinburgh. Whether he did or not, the inspiration that Soseki clearly found in Geddes's writing significantly deepens our sense of Geddes's international influence around 1900.[48]

Wherever they originated, these early twentieth-century thinkers shared an intellectual project of cultural nationalism as a basis for international co-operation. This is expressed not only in the work of Geddes, Nivedita and Coomaraswamy but also in the work of the European art historian Max Dvořák, who was coming into his own as a member of the Vienna School at this time. Mitchell Schwarzer has pointed out that a reading of Dvořák leads one towards 'recognizing and representing artistic and cultural difference'.[49] Schwarzer comments that 'unlike conservative German nationalists of the nineteenth century, Dvořák turned to nationalism as a means of critical resistance to classical norms'.[50] One could write of Nivedita or Geddes in similar terms. Dvořák's words also give a wider context to the invitation to Charles Rennie Mackintosh and Margaret Macdonald to exhibit at the Vienna Secession exhibition of 1900, for what characterised both the art of the Glasgow designers and the art of the Vienna Secession was the exploration of cultural nationalist material as part of a progressive international movement. Typifying that cultural revivalist modernism is Mackintosh's radically modernist square composition dating from about 1901, *There is hope in honest error none in the icy perfection of the mere stylist*. The quotation is from J. D. Sedding and is presented entirely in rectangular letters, yet there are also visual 'signatures' referencing the clans of Macdonald and Mackintosh, through the inclusion of formalised images of the Highland plants emblematic of those clans, the heather and the blaeberry.[51] Such traditionalist modernism has an immediate resonance with Max Dvořák's thinking, which developed in the wake of the Vienna Secession.[52] Like Patrick Geddes, Dvořák recognised in nationalism a device for the remaking of cultural pluralism, as distinct from a device for insisting on cultural homogeneity.[53] It is of interest to identify a wider grouping of these myth-and legend-driven modernists. Many were born in the 1870s. One can include – along with Dvořák (b. 1874) – Ananda Coomaraswamy (b. 1877), C. G. Jung (b. 1875), Nicolas Roerich (b. 1874), J. D. Fergusson (b. 1871) and Abanindranath Tagore (b. 1871). Slightly older are Charles Rennie Mackintosh (b. 1868), Sister Nivedita (b. 1867), Nasume Soseki (b. 1867), W. A. Craigie (b. 1867), John Duncan (b. 1866), W. B. Yeats (b. 1865), Margaret Macdonald (b. 1864), Kakuzo Okakura (b. 1863), Swami Vivekananda (b. 1863), Gustav Klimt (b. 1862) and Rabindranath Tagore (b. 1861). In the anchor roles are Geddes himself (b. 1854), James Frazer (b. 1854), Sigmund Freud (b. 1856) and W. R. Lethaby (b. 1857), and grounding them still further in the analysis and exegesis of myth are Annie Besant (b. 1847), William Robertson Smith (b. 1846) and Andrew Lang (b. 1844).

Geddes may have been unsuccessful in persuading the city authorities of Paris to preserve the *Rue des Nations*, but his involvement in the Paris exhibition of 1900 was a crucial moment in his career. The European dimension of his work had always been strong, but through his friendship with Sister Nivedita in particular he would now find himself engaged in cultural revival as a global phenomenon.

10

City Development as Interdisciplinary Project

THE DUNFERMLINE PLAN: INSTITUTES OF SCIENCE, HISTORY AND ART

THE YEARS AFTER 1900 shed light on Geddes's interdisciplinarity with respect to two disciplines of which he was a pioneer – sociology and town planning. In 1904 Martin White wrote to Geddes not in Dundee or in Edinburgh but at a new address, Low Valleyfield, Dunfermline. Geddes was now fifty years old, and this was his temporary residence while he was engaged in researching and writing *A Study in City Development: Parks, Gardens and Culture-Institutes*. That report would become recognised as a classic statement of the principles of urban conservation and renewal.[1] It was Geddes's proposed plan for the area around Pittencrieff Park in Dunfermline, which he had undertaken in response to a request for proposals from the Carnegie Dunfermline Trust. But by the time Geddes was ready to publish his copiously illustrated book of over two-hundred pages, White appeared distinctly unenthusiastic about reading yet another ideas-packed document by his old friend. He wrote from London on 15 June 1904: 'I shall be glad to see the Dunfermline report but hope it is not too long.'[2] Too long it may well have been for White, but it is nevertheless one of the most significant documents in the history of urban planning. For Geddes, cities meant communities and communities meant culture: not in the sense of an isolated high culture, but in the sense of that continuum of human activity which extends from the child entranced by a pebble to the musician playing the most intricate counterpoint. For Geddes that culture must be expressed through both arts and sciences: the beauty of the pebble *and* its geology, the music *and* its mathematics. On the title page of the Dunfermline report, Geddes extends the interdisciplinary field, emphasising his role as both sociologist and botanist, reminding the reader from the outset that the culture of our cities is founded on our dependence on plants.[3] Thus parks and gardens are not a kind of decorative element in an urban plan, they are links within the city to the wider ecology of the planet, our route of return to nature, while the museums and culture institutes within those parks and gardens enable us through the provision of information and our reflection upon it to understand our human condition – itself part of nature.

It is an irony that White's enthusiasm for his friend's schemes was showing signs of waning at the very time when his consistent financial support for Geddes was beginning to bear such notable fruit. The essential creativity of the relationship between the two men is indicated by the fact that thanks in large part to Geddes's inspiration as far back as 1881, White was increasingly conscious of the possibilities of the new discipline of sociology, and was by this time firmly committed as a patron to establishing sociology on the academic map. Geddes was still very much involved in that project but White was beginning to view the subject with a different emphasis. Although both were involved in the foundation of the Sociological Society, White criticised a circular by Geddes relating to the foundation of the society as 'too purely scientific'. Instead he advocated more emphasis on philosophy, and that these differing emphases should be reflected in two distinct sections of the society.[4] Where once White might have deferred to Geddes on academic matters, it is clear that he no longer did so. At the end of the letter White introduced a topic which was to have major ramifications: he mentions his anxiety to establish a chair of social and political philosophy at St Andrews University. It seems that nothing came of this scheme in Scotland but it did elsewhere, for White went on to endow the first two chairs of sociology at the University of London.[5] From 1905 he gave money for 'temporary endowments of Teacherships in Sociology, including Ethnology', and on 21 September 1907 he gave £10,000 for the foundation of a permanent chair in sociology, in due course held by L. T. Hobhouse.[6] One can note in the light of White's injunction to Geddes in 1903 to pay more heed to philosophy that Hobhouse was by training a philosopher. In 1911 White made funds available for a second chair, to which Edward Westermarck was appointed. It is, perhaps, something of a surprise to realise that the University of London owes its early pre-eminence in social science at least in part to wealth created by Dundee industry. But the debt is not just to Martin White as a benefactor; it is also to Geddes for his intellectual influence on White.

While White was hoping that the Dunfermline report was not too long, another Dundee figure had contributed to it. John Duncan, now returned from Chicago, had made a number of illustrations for the report, including a *Symbol of Eternity* from which poured the water of a mill-lade, and a number of proposals for the Institute of History which included an allegorical statue, *Time and the Fates: the Dial of History*, and other statues of *St Columba* and *Ossian*.[7] He also contributed a mural design of one of the key events in the history of Dunfermline and indeed of Scotland, *The Marriage of Malcolm and Margaret*.[8] In his text Geddes invokes Duncan's previous mural work at Ramsay Lodge in Edinburgh to demonstrate the practicality of revitalising cultural history through the visual arts.[9] For context he draws attention to other works, first of all a frieze of Scottish history which W. G. Burn Murdoch had designed. It was already available for educational purposes in paper form but was 'awaiting fitting realisation in sculpture, in colour, and in actual periodic pageant'.[10] Secondly he draws attention to the then recently completed murals of historical figures and scenes painted by William Hole for the Scottish National Portrait Gallery, and thirdly he notes the historic paintings commissioned for Glasgow City Chambers 'from four of its most eminent artists'.[11] The Institute of History was just one part of a network of institutes and museums set within proposed parks and gardens (both wild and structured) which would have provided Dunfermline with a community university that would have enabled participation at all levels from kindergarten to professorial. In Tom Hubbard's words, 'Dunfermline was to be an all-round experience, a palpable, physical expression of Geddes's desire to counter the fragmentation and specialisation of knowledge.'[12] It was an exemplary approach which still inspires today.

The report had its impact on twentieth-century thinking not least through Lewis Mumford; indeed, it inspired him to embark on his life's work when he first discovered it in 1915.[13] Later he illuminated Geddes's interdisciplinarity when he wrote that 'Olympus and Parnassus were as real to Geddes as the primeval slime out of which the protozoa had emerged'.[14] The mention of Parnassus, the home of the Greek muses, reminds us that Geddes was acutely conscious of the antecedents of the word 'museum'. Thus when he writes of museums in the Dunfermline report he is writing of places which draw their strength from the integrated body of interdisciplinary cultural knowledge which was represented in ancient Greece by the nine muses. Such muse-driven culture institutes are psychologically central to Geddes's planned gardens and parkland. Matthew Jarron has noted that 'for Geddes museums should always be at the heart of town planning', and that what Geddes proposed for Dunfermline was 'to turn Pittencrieff park into an open air museum with reconstructions of human habitation through the ages'.[15] An open-air museum had opened at Skansen in Sweden over a decade before the Dunfermline report was published, but in Scotland such ideas had to wait for their expression until the 1930s.[16] In Geddes's mind such approaches required the research underpinning of a generalist sociology that enabled a fuller understanding of how societies function. To that end, with Victor Branford and Martin White he helped to found the Sociological Society in 1903; that is to say at exactly the time that the Dunfermline report was being prepared.[17] It was to the society's base at Le Play House near Waterloo Station in London that Lewis Mumford was to come in 1920.

Urban planning for Geddes was thus the process of bringing human culture and the natural environment into appropriate relationship through 'the central and significant fact about the city', namely that it 'functions as the specialized organ for social transmission',[18] the nature of the city being to accumulate and embody 'the heritage of a region' and combine 'in some measure and kind with the cultural heritage of larger units, national, racial, religious, human'. Geddes goes on to note 'the individuality of the city – the sign manual of its regional life and record', contrasting that with the wider 'marks of the civilization, in which each particular city is a constituent element'.[19]

This view of the city in context draws one back to ancient Greek philosophy, in particular that of Aristotle, who Geddes described as 'the founder of civic studies', and who 'urged that our view be truly *synoptic* . . . a seeing of the city and this as a whole' and understood the city 'from hill-top and from sea'.[20] Sabine Kraus has noted that 'for Geddes, this synoptic vision is not only that of the city, but also that of nature'.[21] The Dunfermline report was a developed statement of that concise, human ecological view. It led directly to Geddes's seminal exhibitions and planning activities in – as well as Edinburgh – London, Ghent, Dublin, India, Palestine and France. In *Cities in Evolution* Geddes writes a chapter entitled 'Ways to the Neotechnic City'. This idea of 'neotechnic' corresponds to what later commentators have called 'environmentally friendly' or 'sustainable' technologies. The message is clear: cities evolve and we must ensure that they evolve in sustainable ways. That may seem prophetic today but Geddes would simply have seen it as an awareness of the past allowing one to properly understand potential opportunities and difficulties.

CITIES IN EVOLUTION

The significance of the Dunfermline report is hard to overestimate. The report was the first of Geddes's published works to impact on the developing discipline of town planning and it leads directly to his book *Cities in Evolution*. A crucial aspect of the book was to draw attention to one of Geddes's great teaching tools developed at the Outlook Tower, the cities exhibition, and the chapter detailing its history – 'Town Planning and Civic Exhibitions' – occupies almost fifty pages.[22] The book weaves the history of ideas and thinkers into the fabric of the evolving city. By the time of its publication Geddes had established a base at the reconstructed Crosby Hall in the district of Chelsea in London. When he writes of Chelsea he writes, as we have seen, of the educator Jan Amos Comenius, who is for Geddes the interdisciplinary intellectual presence close to Crosby Hall at Moravian Corner at the end of King's Road. For Geddes it is 'but a step in thought from the quiet little Moravian meeting house with its austere cemetery, to one of the greatest and best Puritan movements in history'.[23] Another example of that weaving of thinkers into the urban fabric is Geddes's reference in *Cities in Evolution* to Charles Rennie Mackintosh. He refers to the fact that 'Mackintosh' had become a descriptive term in architecture in Europe.[24] Writing only a few years after the completion of Glasgow School of Art, Geddes already understood Mackintosh's leadership position in the modern movement in European architecture, a position that would not be properly elaborated until Nikolaus Pevsner's *Pioneers of the Modern Movement* was published in 1936. In an earlier unpublished paper Geddes had written, with notable foresight, that 'the real artist is he who, like Mackintosh in the Art College of Glasgow (one of the most important buildings in Europe) gets his effects within the sternest acceptances of modern conditions. For here never was concrete more concrete, steel more steely, and so on.'[25] It is not surprising that a closeness developed between the circles of Mackintosh and Geddes. For example, in 1912 or a little later one finds Mackintosh's wife Margaret Macdonald writing to Patrick Geddes's wife, Anna Morton, lamenting the fact that they have no one like Geddes to give lectures in Glasgow. She goes on to mention visiting the recently married John Duncan and his wife in their new house.[26] Volker Welter has noted that Geddes organised a tour of the Glasgow School of Art building, conducted by Mackintosh, during the International Exhibition in Glasgow in 1901.[27] The visit was under the auspices of the International Assembly, which Geddes had initiated for the Paris exhibition the previous year.[28]

While Geddes's response to Mackintosh shows his understanding of the dynamics of modernism, his absolute commitment to the value of the past is equally clear in *Cities in Evolution*, not least with respect to mention of Crosby Hall in Chelsea.[29] Geddes had entirely reconstructed this historically important London building on a new site, and it was there that his cities exhibition was first shown in February 1911. It marked the formal opening of the hall by John Burns MP, the leading parliamentary influence on the pioneering town-planning legislation enacted in 1909. The Crosby Hall project is yet another example of Geddes's extraordinary energy and it further demonstrates his talent for creative solutions to conservation issues. Crosby Hall had been a building of historic importance in the City of London, sometime residence of Sir Thomas More. Despite public opposition, it had been demolished to make way for bank premises. Care had, however, been taken to record and number the stonework and timbers during the process of demolition. In 1908, despite competition from other bidders, Geddes persuaded the London County Council to give him the bits for nothing on condition that he would re-erect the hall near an area known as More's Garden in Chelsea, close to which he had already held a London summer meeting and established a hall of residence associated with the University of London.[30] Geddes's powers of persuasion extended to finding enough donors and enough expertise to re-erect the hall as a major part of a University Hall for London, echoing that established in and around Ramsay Garden in Edinburgh a quarter of a century before. Crosby Hall thus became an act of building preservation, a student resource, a performance centre and an exhibition space.[31] In setting up the Cities and Town Planning Exhibition at Crosby Hall, Geddes had a notable helper in his fellow Scot Thomas Adams, who would become in due course one of North America's most distinguished planning pioneers. In late 1913 Adams became the first president of the Town Planning Institute in London, with Geddes as honorary librarian.

The background to the exhibition was the work undertaken at the Outlook Tower since its inception. This had led to Geddes's seminal Edinburgh Room for the 1910 Town Planning Exhibition at Burlington House in London. It was in this context that Patrick Abercrombie described Geddes as 'a most unsettling person', using the word 'unsettling' to make clear Geddes's continual questioning of received wisdom and the engaged manner in which he introduced the multiplicity of sources of information that he considered necessary to the planner's task.[32] Geddes's Scottish interdisciplinary background stood him in good stead here, and one should bear in mind that much of his thinking was devoted to designing systems to handle and relate such diverse knowledges. This is what the Outlook Tower was for and this is why Geddes referred to it as an index of indexes. Crosby Hall was the first of many venues for the cities exhibition over the next few years. Next was an invitation from Edinburgh corporation, which resulted in a three-week showing at the Royal Scottish Academy.[33] A number of the images on show would have been the work of the young Frank Mears, who went on to become president of the Academy in 1944. This showing of the exhibition has a further interest for it was opened by Lord Pentland, who in 1912 gave up his role as Secretary of State for Scotland and took up a new role as Governor of Madras. In due course he would invite Geddes to work with him in India. Furthermore, Pentland's father-in-law was Lord Aberdeen, who had recently taken up the role of Lord Lieutenant of Ireland for the second time. Dublin and Belfast were the next venues for Geddes's exhibition. The influential voice in getting Geddes across the Irish sea was the remarkable Ishbel Maria Majoribanks, Lady Aberdeen.[34] She had performed a leading role in the development and maintenance of traditional arts and crafts in Ireland since the late nineteenth century. It was her organisation, The Women's National Health Association, that had invited Geddes in the first instance.[35] The cultural significance of this Scottish-Irish milieu has been explored by Nicola Gordon Bowe and Elizabeth Cumming.[36] Geddes's exhibition led to the foundation of the Housing and Town Planning Association of Ireland. The next invitation was to Belgium, where the cities exhibition was a key feature of the Ghent International Exhibition of 1913, of which Geddes's friend Paul Otlet was one of the organisers. Geddes had high praise for the event. He wrote that it was 'in various respects one of the best thought out and most vitally executed International Exhibitions since that of Paris in 1900, and of the most distinctly civic character'.[37] Philip Boardman has given a vivid account of the order within confusion which characterised Geddes and his supporters at work.[38] Crucial were his wife Anna, his son Alasdair and his assistant Amelia Defries. They may at times have been driven to distraction, but the effectiveness of Geddes's emphasis on creative thinking over received wisdom is reflected by the fact that his exhibition won the Grand Prix at Ghent.[39]

A further connection with Thomas Adams can be noted here. While he was in Ghent, Geddes wrote an introduction for *The Story of Edinburgh Castle*.[40] The author was Louis Weirter, who was the son of a music teacher from Germany who had settled in Edinburgh. Weirter's sister Caroline had married Thomas Adams in 1897.[41] Another link is that the endpapers of Weirter's book were the work of Otto Schlapp, Professor of German at the University of Edinburgh, and a related design had appeared the previous year in Geddes's civic review, *The Blue Blanket*. After Ghent, Geddes was invited to show the exhibition in Dublin, and at this juncture one of his assistants was H. J. Fleure. The experience was a key influence for Fleure, who later adapted Geddes's ideas, referring to 'human regions'.[42] As professor at University College Aberystwyth, he shared Geddes's awareness of cultural nationalism, and this awareness was no doubt sharpened by the politically charged atmosphere of Dublin in 1914. One of Geddes's notable meetings in Ireland was with the redoubtable leader of the Dublin dockers, Jim Larkin.[43]

In 1914 the Town Planning Institute in London showed interest in acquiring Geddes's cities exhibition.[44] Geddes instead decided to transport the exhibition to India, where he was to work at the invitation of Lord Pentland. But by the time the exhibition was in transit on the steamship *Clan Grant*, the First World War had broken out and one of the early casualties was the cities exhibition, sunk by a German warship in the Indian Ocean. *Cities in Evolution* was almost ready for the press but Geddes had time to add one more sentence to the preface: 'The Cities and Town Planning Exhibition, of which so much has been said in the following pages, has fully shared in the civic history it illustrated, by total destruction by the vigilant and enterprising *Emden*, but is none the less in the process of renewal.'[45] And renewed, in due course, it was. Geddes's intellectual resilience could hardly be better demonstrated.

11

Ecological Research in Dundee

ECOLOGY AS A SCIENCE: GEDDES'S STUDENTS AND BARTHOLOMEW'S MAPPING

PARALLEL TO PLANNING WAS botany. For Geddes they were complementary activities and the same period that saw him develop his planning insights in the Dunfermline report saw botanical research in Dundee flourish under his leadership. In the prefatory note to the *Evergreen: Book of Autumn*, published in 1895, Patrick Geddes and Victor Branford wrote of the 'Return to Nature' as 'a rallying call which each age must answer in its own way'.[1] This rallying call underpinned Geddes's planning ideas but it found its deepest research expression in Dundee botanical research as it helped to bring into being the discipline of ecology. But Geddes's growing involvement with planning on both a national and international stage left him open to criticisms of neglecting his role as professor of botany at University College Dundee. Martin White wrote to him in 1904 of hearing 'various rumblings at the little attention Dundee gets from you'.[2] Yet by the time of White's letter Geddes's actual achievement in Dundee was both substantial and pioneering. Here Geddes the sociologist/planner became Geddes the human ecologist. Here he inspired ecological research of the very highest quality from two students in particular, the short-lived Robert Smith and his elder brother William. The focus of this research was the mapping of the vegetation of Scotland, region by region. The complementarity of such mapping with Geddes's regional planning concerns is obvious; the Dundee research can be seen as one of the driving forces of his wider activities. Its background lay in his strong links with Charles Flahault, who had developed a geographical and sociological approach to botany at the University of Montpellier.[3]

Geddes's student William Smith went on to become one of the founders of ecology as a recognised discipline. The exceptional quality of his and his brother Robert's pioneering work was recognised in the first words of J. H. Burnett's definitive *Vegetation of Scotland*: 'The study of vegetation in Scotland was initiated by the brilliant work of Robert Smith, his elder brother W. G. Smith and their collaborators at the end of the nineteenth century.'[4] Burnett goes on to point out that 'University College, Dundee, differed from the other Scottish Universities in having two men of genius and independent thought in charge of its biological departments, Patrick Geddes in botany and D'Arcy Thompson in zoology. Robert Smith . . . entered the college in 1892 as a part-time student, and finally became a lecturer in 1899.'[5] Burnett notes that Smith's interests were focused by a visit, arranged through Geddes, to Flahault's department at Montpellier, where he was introduced to the French school of phytogeography. In his history of the British Ecological Society, John Sheail notes that it was Geddes's practice to encourage his most able students to spend a period under Flahault. Robert Smith accompanied Flahault on numerous surveying expeditions during the winter of 1896–7.[6] A further part of this rich research mix is noted by Burnett, namely that Smith 'returned in the same year that Harvie Brown and J. G. Bartholomew published their 10 miles to 1 inch (1:625,000) "Naturalist's Map of Scotland" '.[7]

Alexander Mather has characterised the period as 'the golden age of vegetation mapping in Scotland' and attributed it to 'a remarkable coincidence of several factors', namely 'the inspirational genius of Geddes and his European connections, the industry and enthusiasm of Robert Smith, [and] the cartographic innovations of John Bartholomew'.[8] And crucially, publication was facilitated by the Scottish Geographical Society in its magazine. Mather goes on to suggest that the vegetation survey project 'can be seen as a metaphor for, and flagship of, the greater Geddes project', its importance being that in the years around 1900 'there was a real prospect that a distinctive Geddesian geography could emerge that was international in outlook, conceptually aware but empirically grounded, holistic, and practical in orientation'.[9] Mather makes clear the historical significance of this lost opportunity, for this was a period that was critical in the evolution of geography as a discipline: 'the significance of the enterprise could have been profound and fundamental: in essence Geddesian geography was human ecology'.[10] One can note that Heinz Maus, in his history of sociology, credits Geddes as the first to apply the term 'ecology' to social phenomena.[11]

But the enterprise was not sustained, for despite the appointment of a talented Montpellier-trained assistant – Marcel Hardy – there were

insufficient resources available to Geddes to support a research community of botanist-geographers in Scotland. The departure from Edinburgh
of Geddes's student and fervent admirer A. J. Herbertson to work with
H. J. Mackinder at Oxford in 1899 was symptomatic.[12] Robert Smith's
death in 1900 made things worse. But what is clear with hindsight is the
quality of research made possible by Geddes and its wider influence on
the development of geography as a discipline. This can be followed in the
frequent references to Geddes in *The Geographical Tradition: Episodes
in the History of a Contested Enterprise* by David N. Livingstone, who
also notes the significance of Geddes's thinking for another pioneer of
the teaching of geography, H. J. Fleure, who in 1918 published his first
book, *Human Geography in Western Europe*, in Geddes's *Makers of the
Future* series.[13] As already noted, Fleure had assisted Geddes in Dublin
in 1914 and Iain Stevenson credits him as the major figure in 'the making
of the Geddesian tradition in geography'.[14] Geddes's influence is further
clarified when one considers the early history of ecology as a recognised
discipline. The work of both the Smith brothers was fundamental. The
surviving brother, William, is credited, along with A. G. Tansley, as one
of the two founders in 1904 of a key organisation, the British Vegetation
Committee. One of its active members was Marcel Hardy. By 1913
the committee had outgrown its original remit and developed into the
British Ecological Society, the first ecological society in the world.[15]

Also notable is the encouragement and expert advice given to the
brothers Smith by the mapmaker John G. Bartholomew.[16] The importance of Bartholomew's contribution to their projects is hard to overestimate, and it leads one to note the high production and aesthetic
values of the *Scottish Geographical Magazine* in which these vegetation
maps were published.[17] In 1900 Robert Smith published the first two
parts of his *Botanical Survey of Scotland*, those for Edinburgh and for
North Perthshire.[18] It is sad to reflect that the same volume also includes
Smith's obituary, written by Geddes himself.[19] Further valuable papers
were written by William Smith.[20] Geddes maintained close links with the
magazine and in 1905 he contributed a substantial two-part obituary of
Élisée Reclus.[21] In the same volume is William Smith's botanical survey
of Forfar and Fife.

The irony of White's implied criticism in 1904 could hardly be clearer,
for Geddes was in the process of giving University College Dundee
research leadership in the nascent science of ecology. But as was so
often the case, he was facilitating work in a new discipline which had
not as yet been recognised as such: as a result, his own contribution was
not perceived at the time. B. T. Robson sums it up when he notes that

Geddes 'paid the price of pioneering a multidisciplinary approach at a stage when the disciplines themselves were unformed'.[22]

Geddes's planning insights cannot be considered in isolation from his leadership of ecological research.[23] This is symbolised by the fact that the time of the publication of the Dunfermline report is also the time of the publication of the regional vegetation maps in *The Scottish Geographical Magazine*. This must be borne in mind when considering Geddes's influence on the developing discipline of town planning in Britain. It is more than a pity that Geddes was never able to bring his ecological research together with his planning ideas in Dundee. He made strenuous attempts to do so, particularly in his attempt to create a botanical garden for the city at the eastern end of Magdalen Green. Its significance was as the heart of a wider plan for integrating various garden schemes to create a green corridor from Magdalen Green to the heart of the university.[24] The scheme failed to gain council approval by the narrowest of margins in 1909. Geddes had made substantial reference to Dundee in the Dunfermline report, and had that botanic garden scheme gone ahead, consideration of Dundee would no doubt have been prominent in his book *Cities in Evolution*.[25] As it is, the city gets little mention, although in one passage of illuminating comparisons Geddes makes clear his awareness of the post-industrial challenges that the Dundee would face:

> Edinburgh is not permanently destined to professional fossilisation, legal and other; Dundee need not accept ruin by Oriental competition at the lowest level of subsistence; Dublin will not further subside into squalor, nor Belfast into bitterness; but each and all revive, through fuller appreciation of their respective possibilities and cultivation of their advantages, and towards completer and higher inter-civic co-operation.[26]

THE BRITISH ASSOCIATION MEETING OF 1912

Perhaps there is an echo of what might have resulted from a co-operation between academic botany and urban planning in Dundee in the title of the paper Geddes presented in September 1912 to the meeting of the British Association in Dundee: 'Regional and Civic Surveys: the needed Co-operation of the Sciences towards the Town Planning Movement'. Geddes, reflecting his growing role as a sociologist, was a vice-president of Section F of the Association meeting, 'Economic Science and Statistics'. Another contributor to the section was Ramsay MacDonald MP, speaking on 'The Minimum Wage'. Geddes's former Dundee student William Smith, building on his botanical survey work, gave a paper (jointly with C. B. Crampton) on 'The Influence of the Origin and Topography of Grasslands'. This was presented to the newly formed Section M (Agriculture), which had developed out of Section F.[27] One of Geddes's former Edinburgh students also gave a paper. This was the polar explorer William Speirs Bruce who presented to Section D (Zoology), of which D'Arcy Thompson was a vice-president.[28] Bruce spoke on 'Zoological Results of the Scottish National Antarctic Expedition'. Along with Geddes, Thompson had supported Bruce in his ventures, and despite their respective commitments to other sections it is possible they may have been present to hear Bruce give another presentation in the session of Section E (Geography) devoted that year to Antarctica. Something of the drama of the development of geography emerges here. In the official report of the meeting published a year later, it is notable that Bruce is not even listed as a contributor to this section. This gives a clue to the turbulent state of polar research during the period. A vice-president of Section E was Sir Clements Markham, who contributed a paper on 'Antarctic Discovery' that was noted in the proceedings as opening a discussion on the Antarctic. The full discussion is reported in the *Geographical Journal*, and there one finds that Markham's paper was directly followed by a shorter but nevertheless substantial contribution by Bruce.[29] Reported also is discussion from Bruce's supporter (and later biographer) Rudmose Brown, who had been Geddes's assistant in Dundee from 1900 to 1902. In latter years Markham's character has become a subject of some interest.[30] His attempts to thwart Bruce's careful scientific explorations in favour of the more heroic style of his own protégé, Robert Falcon Scott, have been noted by Peter Speak.[31] The fact that Bruce's paper to the British Association in Dundee is not recorded in the proceedings is consistent with these attempts. The tragedy of Markham's contribution is that part of his agenda seems to be to make of Scott a legend in his own lifetime, yet by the date of the Dundee meeting, unknown to Markham, Scott had lain dead and undiscovered in the Antarctic for several months. Markham's approach has been characterised as 'freighted with a nostalgia that got out of hand, a reverie trying to be real'.[32] An irony is that in his selective account of Antarctic exploration, Markham not only ignores Bruce, he also makes no mention of the Norwegian Roald Amundsen, who had by this time reached the South Pole in advance of Scott and survived the journey. By contrast, in the discussion reported in the *Geographical Journal*, Rudmose Brown makes sure that Amundsen's previous explorations are given due credit.[33] A final critique of Markham comes from another discussant, T. V. Hodgson, who describes Markham's advocacy of human rather than animal traction in polar exploration as 'a serious mistake'.[34] Whether Scott's death can be attributed to that 'serious mistake' is a moot point but the comment illustrates the clash of ideologies that surrounded Antarctic research. For many years such controversy was obscured by the myth of Scott that Markham so assiduously created, but thanks to the work of Peter Speak and Geoffrey Swinney, Bruce's contribution can now be appreciated.[35] Although Bruce's work may not have been given proper credit by Markham, a substantial number of his photographs of the Antarctic had been already published in a popular format in 1907 by the Glasgow publisher Gowans and Gray.[36]

Martin White had no obvious role in the British Association meeting but he had written to Geddes on 6 September hoping to see him there.[37] He did, however, make a significant contribution to the event, for he had presented John Duncan's painting *The Riders of the Sidhe* to Dundee Corporation in 1912 and it was exhibited as part of a major exhibition mounted in the Victoria Galleries in Dundee to mark the meeting.[38] *Riders of the Sidhe* had been shown at the Royal Scottish Academy in Edinburgh the previous year, and it shows Duncan at the height of his powers as a Celtic revival symbolist. It is one of Duncan's most important pictures and its significance for Geddes's milieu is discussed in more detail in chapters twelve and thirteen.[39] The Dundee exhibition was notable both for its scale and its quality. International in scope, it had Scottish art at its core, including a large number of works by Raeburn. Among much else, there was a previously unexhibited late work by the pioneer of modern painting in Scotland, William McTaggart, who had died in 1910.[40] Through the patronage of McTaggart's friend the Dundee industrialist James Guthrie Orchar, Dundee had developed a close link with the artist. D'Arcy Thompson was on the exhibition organising committee, as was Geddes's old friend R. C. Buist.[41] Geddes would have seen the exhibition as a natural complement to the scientific work of the British Association.

12

Dramatising the Past and Informing the Future

MASQUES OF LEARNING

THE SAME YEAR THAT Geddes spoke at the British Association meeting in Dundee, the project at the heart of his early initiatives in Edinburgh celebrated twenty-five years of activity. Geddes used the semi-jubilee of University Hall in 1912 to drive a major civic education project which brought together his thinking in dramatic form: he called it *The Masque of Learning*. Whereas the cities exhibition reflected on the culture of cities through their form, the masque engaged directly with civic history through performance. The extent of the masque was global but – as with *The Evergreen* – its interdisciplinary internationalism was rooted in Scotland. For Geddes, drama was a means to connect city and citizenry using history and geography to illuminate the development of ideas and vice versa. While the high point of these dramatisations of history came in 1912 and 1913, as Edward McGegan recalls, Geddes had produced small-scale masques and pageants as early as 1895.[1] But the 1912 production in Edinburgh was on a different level, bringing together some six-hundred and fifty participants. It attracted numerous letters of support and congratulation.[2] The publication that accompanied it had the crest and motto of University Hall inside its front cover and nothing could better exemplify that motto, *Vivendo discimus*, than the experience of the masque participants. [Figure 9.] It also had a frontispiece by Stanley Cursiter, then still in his twenties but later to become the keeper of the National Gallery of Scotland.[3] Although best known as a landscape painter, Cursiter had a long-standing interest in book illustration, some of it of a Celtic revival nature.[4] This gives context not only to his involvement with Geddes but also to John Duncan's exhibition, the first by a living artist, held at the National Gallery of Scotland in 1941, during Cursiter's period as keeper. At the time he was working with Geddes, Cursiter was also experimenting with Futurist compositions, and his exploration of such modernist ideas bore fruit in his later advocacy of a Scottish National Gallery of Modern Art.[5]

Insight into the development of Geddes's masques as part of a wider cultural movement can be gleaned from the involvement of many of Geddes's family and associates in the organisation and performance of the Scottish pageant and masque at the Scottish National Exhibition in Saughton Park in Edinburgh in 1908.[6] Geddes himself was not a performer, no doubt because he was at that time campaigning to rebuild Crosby Hall in London. In its reconstructed form Crosby Hall was to be the London venue for Geddes's masques in 1913. In 1908, as *The Scotsman* reported, Geddes's son Alasdair and his daughter Norah both took part in the 'Celtic Races' part of the masque, which was organised by John Duncan. It shared many features with the Celtic part of *The Masque of Learning* as developed in 1912 and 1913.[7] Others involved included the young Helen Schlapp as an attendant of St Bride; Duncan painted her on the island of Iona the same year.[8] She was the daughter of Otto Schlapp, at that time lecturer (subsequently professor) in the department of German at the University of Edinburgh.[9] She would later marry H. S. (Jim) Ede, who founded Kettle's Yard gallery in Cambridge. The part of the Celtic deity Lugh was performed by 'Mr A. F. Whyte'. Aird Whyte was the son of Jane and Alexander Whyte, and brother of the pioneering biologist Lancelot Law Whyte. As I noted in chapter eight, Geddes had advised Jane Whyte with respect to Aird's education in 189.[10] Among the Riders of the Sidhe were the painter Eric Robertson and Geddes's son Alasdair.[11] Geddes's daughter, Norah, was one of the *sidhe* (faeries). John Duncan himself took the part of Cormac, High King of Ireland and legendary ancestor of Clan Donald. Painters were well represented: for example, the part of Queen Maeve was taken by Katherine Cameron. She was the sister of D. Y. Cameron, and is an interesting figure to note here, not least for her role as an illustrator of children's stories, some of which were written in Gaelic.[12] She is also a link to the milieu of Charles Rennie Mackintosh and thus provides a reminder of the significance of the Edinburgh aspect of the Celtic revival for Mackintosh's Glasgow.[13] The year before the 1908 pageant, Margaret Macdonald Mackintosh and her Glasgow School of Art colleague Ann Macbeth had contributed to a Celtic revival miscellany edited by Elma Story, a painter with very close links to Iona.[14] Containing literary contributions in both Gaelic and English, *Am Bolg Solair/The Pedlar's Pack* was published by the Celtic Press in Glasgow.

D. Y. Cameron was also a contributor to that publication and three of the other artists involved – Charles Mackie, William Walls and James Cadenhead – had been involved in *The Evergreen* a decade before. In the 1908 pageant, Cadenhead took the part of Fionn (Fingal). Another notable painter, Edwin Alexander, took the part of Ossian's son, Oscar. Alexander Carmichael's son Eoghan, who would later write a short book on Celtic art, was Diarmid.[15] Among the Vikings was the sculptor William Grant Stevenson (brother of David Watson Stevenson), whose works include statues of William Wallace in Aberdeen and of Robert Burns in Kilmarnock. Another Scottish sculptor, Harry Gamley, figured as Cuchulainn.[16] John Duncan's 'Celtic' section was thus really a Gaelic legend section, drawing (according to *The Scotsman*) on two works in particular, H. d'Arbois de Jubainville's *The Irish Mythological Cycle* and John Francis Campbell's *Leabhar na Feinne*. The reference to Campbell is particularly interesting in view of his crucial place as a collector of Gaelic tales in the second half of the nineteenth century. It should also be noted that Campbell was one of the first authors to adopt the term 'Celtic art', when he used it in 1862 in volume four of his *Popular Tales of the West Highlands*.[17] Duncan's section was complemented by another Celtic aspect of the pageant, 'Arthurian Legend', described as 'Glasgow's contribution to the pageant'. John Duncan's counterpart as organiser was the outstanding graphic artist Jessie M. King. Her sources included both Malory's *Morte d'Arthur* and Sebastian Evans's then recent translation *The High History of the Holy Graal*. A few years earlier, King had drawn a notable set of illustrations for the latter work.[18] Her band of helpers reads like a *Who's Who* of Glasgow School of Art. She herself took the part of the Angel of the Holy Graal, while her husband the painter E. A. Taylor (who would write an article on John Duncan for *The Studio* in 1920) took the part of Sir Galahad. The photographer and pioneer of photogravure, J. Craig Annan, took the part of King Arthur, and it is to Annan that we owe a series of portraits which have become the standard images of the Glasgow artists of the early twentieth century. Jessie Keppie was Enid and Isolde was played by a Miss Raeburn (presumably the painter Agnes Raeburn), while Morgan le Fay was played by Charles Rennie Mackintosh's wife Margaret Macdonald, described in *The Scotsman* as Mrs C. Rennie M'Intosh. The 1908 Edinburgh pageant can thus be seen as a conjunction of Edinburgh and Glasgow Celtic revivals, an indication of the firm links between Geddes's milieu and that of Mackintosh. In *The Masque of Learning*, Geddes describes the 1908 pageant as 'magnificent' and singles out for praise the Arthurian section by the Glasgow artists.[19]

The part of the pageant on the early Church was organised by Phoebe Traquair.

What characterised Geddes's masques over and above the spectacle of a pageant was his commitment to community drama as an educational tool. The spectacle of pageantry was thus transformed into the reflective drama of the masque. For Geddes the history of ideas thus explored was linked to the individual biography of every student in every hall of residence; hence the occasion for the first performance of Geddes's *Masque of Learning* was to celebrate University Hall:

> The Masque of Learning, in its initial form, was prepared to celebrate the twenty-fifth anniversary of the foundation of University Hall, Edinburgh. It was performed – by about 650 active participants, as players, orchestra and choir – in the Synod Hall, Edinburgh, on 14th to 16th March 1912, and repeated on a later evening for pupils of the schools of Edinburgh.[20]

Geddes notes that large numbers failed to obtain admission and 'encouraged by the interest of all ages and classes of the community' those involved, instead of dispersing, formed themselves into an association – 'The Edinburgh Masquers, Outlook Tower'.[21] He then notes 'the addition of a full score of new scenes'. That expansion enabled the development of the original masque into two further ones, *The Masque of Ancient Learning* and *The Masque of Medieval and Modern Learning*. Cursiter's design was adapted for the new publications.[22] Geddes's description continues: 'each Masque is complete within itself and may be visited and understood apart from its fellow', but it is stressed that 'the two make up a more comprehensive presentment than has before been attempted upon any stage, of main phases and salient events in the long and changeful history of civilisation'. Geddes's conclusion is that 'in short, the two present Masques thus embody an outline of the World-Pageant – of Education, of Learning, of Culture, of Human Advance'.[23] The masques moved on to London in 1913 to be performed by, in Geddes's words, a kindred group of some '500 London Masquers, at Crosby Hall'.[24]

Geddes's masques prefigure the educational aims of H. G. Wells's *Outline of History*, which began publication in parts in 1919. In line with his sociological and evolutionary thinking, Geddes uses the earliest scenes (prehistory, Babylon and Assyria) to establish the importance of the family, the city and the consequences of militarism. Via Egypt he turns to geometry as a method of linking cosmos and ecology, and points out that with this method comes the beginning of academic institutions.[25] This sets the scene for (among much else) 'the fullest presentment of the Masque', namely that of ancient Greece. For Geddes, 'first, upon the normal background of nature and labour comes the undying song of Homer, the choric dance of youth, the evocation of Heroes and Gods, of Goddesses and Muses'. He then proceeds to historic Greece, 'with its founders of modern science and medicine, of philosophy and politics, of architecture, sculpture and drama; and, above all, of City and citizenship'.[26] This illustrates the complementarity in Geddes's thought between myth and cultural history, between muses and city, so to speak.[27] Much as was the case with his younger contemporary C. G. Jung, Geddes understood figures from myth as methods of thought, recognising that they represent fundamental aspects of the human psyche.[28] Geddes goes on to consider Rome 'in justice and in peace from Clyde to Euphrates: an inspiring citizenship and humanism in one'.[29] An idealised view to be sure, but one which for Geddes indicated the possibility of a Europe united by shared civic values.[30] Geddes's intention here, as with the stained-glass windows in the Outlook Tower, was to provide starting points for cultural comparison, and as with the Outlook Tower he roots his international message in the familiarity of region and locality: 'It is time next to realise that our own regional heritage, that of the Celtic world, which once stretched from Galatia to Gaul, and on either side beyond, was no mere Barbarian darkness, but had also its life and light.' And turning to the relevance of that history to his present he writes of its survivals, from Brittany and Wales to Ireland and the Scottish Highlands, preserving the embers of their ancient and sacred fires. 'Fairylands and mythic paradises, memories of beauty and heroism, types of chivalry and saintship, live on unforgotten' and ripe for revival to go forth 'in ardent souls . . . to civilise their conquerors, to guide them anew to meditation and learning, to spirituality and poetry, to lead them in art and song'. And he concludes: 'A Celtic Epilogue will therefore end the Ancient Masque.'[31]

ST COLUMBA AS POINT OF REFERENCE

At the heart of that Celtic epilogue was the dramatic scene of erecting a statue of St Columba, at the head of Edinburgh's Lawnmarket, close to the Outlook Tower. Geddes saw the saint as a unifying figure of Scottish past and present. I noted in chapter seven that Geddes used John Duncan's drawing *Columba on the Hill of Angels* as the frontispiece to Victor Branford's 1912 book, *St Columba: A Study of Social Inheritance and Spiritual Development*, a volume that was published by Geddes as a complement to *The Masque of Learning*.[32] In view of my earlier comment with respect to the low-cost, high-print-run nature of the masque texts, one can note that Branford's *St Columba* was published both to match those texts, and a year later in a version more in line with the high production values of the Celtic Library of the 1890s.[33] In both editions the book took as its cover an illustration from a sketch of a proposed statue of the saint made by Percy Portsmouth, professor of sculpture at the newly founded Edinburgh College of Art. This is the statue on which the epilogue of the masque focuses. Branford's book contains the following description of its close link with the project:

> Around this model centres the Epilogue Scene of the Masque of Ancient Learning, which follows upon the commemoration of St Columba and his missions. This Epilogue anticipates the erection of the statue, and suggests a site – a literal 'St Columba's Place' – upon the Historic Mile of Old Edinburgh, which after long dilapidation and neglect, is again in process of conservation and renewal.[34]

Branford proceeds to provide an intriguing intellectual context, in which he emphasises the sociological approach of Geddes, which he describes as the 'Comte-Le-Play-Geddes formulae' that 'resume the sociology of the past two generations'. Then he notes the complementary psychological dimension – what he calls the 'Lange-James-Hall formulae, which, in more empirical fashion, have done a similar service for psychology'. And he continues by drawing attention not only to Geddes's papers published by the Sociological Society, but to Stanley Hall's work on adolescence and William James's *The Varieties of Religious Experience*.[35] The last-mentioned work had its origins in James's series of Gifford Lectures to the University of Edinburgh in 1901 and 1902. The significance given to American psychology alongside European sociology is of considerable interest. Geddes sent a copy of Branford's *Columba*, along with *The Masque of Learning*, to the pioneering American psychologist Stanley Hall in Massachusetts, with a covering letter in which he writes of the masque as 'a growing attempt to visualise the essence of the whole procession of culture, from Prometheus to today, and even tomorrow' and continues, 'since you are good enough to express interest in the Outlook Tower, let me point out that this summarises one of its main aspects – that of historical synthesis'.[36] This underlines again the complementary nature of the tower and its associated activities. Geddes describes the masque as an attempt 'to outline – or at least suggest – the general march of culture, the essentials of social phylogeny', but encourages Hall to 'look also at Branford's St Columba as an example of autogeny, and realise how largely this co-ordinates with your own work', emphasising the complementary roles of the individual and group in the development of social formations. Geddes goes on to acknowledge Hall's achievement in turning 'the attention of the educators and of adults generally to the adolescent', but sees his own role as suggesting 'the complemental task – that of arousing the adolescent to his quest, and thus worthily applying his moral passion and individual genius; and similarly for adult and senescent'.[37] Geddes's interest in both individual and group processes essential to effective education could hardly be clearer.

John Duncan discusses his frontispiece for Branford's book in a letter to Geddes and notes his interest in making further Columba works, including one modelled on his memories of Father Allan McDonald of Eriskay sitting at the helm of a boat.[38] Eriskay was a culturally significant place for the Celtic revival, not least because of Father Allan McDonald's activities as a recorder of oral Gaelic culture. A friend of Alexander Carmichael, Father Allan was an important advocate of the cultural traditions of the Western Highlands. He was also a poet in his own right; indeed, Ronald Black has described him as picking up the torch of Burns and lighting a path into the future towards Sorley MacLean.[39] In 1905, the year of Father Allan's untimely death, Duncan painted him conducting a funeral on the shore at Eriskay.[40] There is another image by Duncan, presumably of the same event, reproduced in Grace Warrack's book *From Isles of the West to Bethlehem*.[41] In Eriskay Duncan was known as *Iain na Tràghad* (John of the Beach) or *Iain a Chladaich* (John of the Shore).[42] This naming is resonant with the Hebridean coastal setting of his 1915 painting, *Adoration of the Magi*.[43]

An indirect reference to St Columba can also be found in the fabric of the Outlook Tower itself. On the parapet is an arrow marked 'Iona'. It puts one in mind of the comment that Geddes would 'cultivate the region, not alone for the sake of the local life but for the sake of the world's life, too: he helped to revive national traditions, not to inflame a "sacred egoism" but to contribute to the wealth of humanity'.[44] Geddes summed up his view of the significance of Columba and Iona when he wrote in his *Masque of Ancient Learning*:

> The recent reopening and present restoration of Iona Cathedral are plainly of far more importance than Scotland has yet commonly realised. In all our stormy history there is but one single figure whom all sections, divisions, classes, faiths of this much divided land have accepted as representative – St Columba.[45]

There had been earlier such explorations in Geddes's milieu. Thus in 1897 J. Arthur Thomson and William Sharp (in the person of Fiona

Macleod) had contributed to *Songs and Tales of St Columba and his Age*.[46] And again in the person of Fiona Macleod, Sharp's extended essay *Iona* published in 1900 had explored the Celtic world. It begins thus: 'A few places in the world are held to be holy, because of the love which consecrates them and the faith which enshrines them. Their names are themselves talismans of spiritual beauty. Of these is Iona.' He goes on to call Iona 'the Mecca of the Gael'.[47] This spiritual geography was given visual form by D. Y. Cameron in a number of landscape works of Iona and the surrounding area which were used as frontispieces for the collected works of Fiona Macleod, issued from 1910 onwards.[48] Cameron, more than any other artist of the time, treated the Highlands and Islands of Scotland as a landscape of spiritual significance. Lindsay Errington has described his work as having an austere Celticism.[49] One can sense this, for example, in his prints and paintings of peaks such Goat Fell on Arran and Ben Ledi in the central Highlands.[50]

Although *The Masque of Ancient Learning* concludes with Celtic Christianity, Celtic legends are equally important. Here Geddes re-establishes and extends the discourses set up in *The Evergreen* and Ramsay Lodge in the 1890s: 'Ossian the Celtic Homer, appears with Malvina.'[51] I have noted the occurrence of these bards in John Duncan's *Anima Celtica*. Even closer in date to the masques is the image by Robert Traill Rose, whose pen-and-ink drawing *Ossian and Malvina* took its place as the frontispiece for Keith Norman MacDonald's book, *In Defence of Ossian*, published in 1906.[52] Geddes notes that in the persons of these two bards, 'the mythology and legend of Druid and Druidess thus give way to poetic art' and that 'as Homer's song evoked the heroines and heroes of the past, so again Ossian's. Hence Cuchullin, the Celtic Achilles, comes back from the shades, and Deirdre, a nobler Helen, returns with the sons of Usnach.'[53]

Every one of these subjects had by then been given visual form by John Duncan. As Geddes continues, the relevance of Duncan's work is even clearer:

> Next, as the poet changes his song from historic epic to hymn of creative idealism, there pass before us the Riders of the Sidhe, each offering one of the four gifts of life to men. First the leafing, flowering, fruiting branch of the Life-Tree – the simple life and labour of the People. The next bears the cup – for the joy of Life in its prosperity. The next is gazing into his magic crystal, of Thought; in which, from reflections from without, again from memories within, Emotion, Reason and Intuition are ever creating new visions. Finally comes the bearer of the Sword-for Idealism in Action, Justice in Rule.[54]

This is the precise subject of Duncan's painting *The Riders of the Sidhe*, exhibited at the Royal Scottish Academy in 1911 and published as a finished drawing in *The Blue Blanket* in 1912.[55] In a way which relates directly to his diagrammatic notation of life, Geddes continues, 'these four essential types are demonstrable in the psychology as well as the economics and politics of our modern towns'. He sees these types as expressed in the urban fabric itself: 'the Town of industry, the School of learning, the Cloister of religion, thought or art, and in the Acropolis Temple or Cathedral, of fully realised culture and citizenship'. And he sees these as the 'four essentials of every City worth the name'.[56]

'Within the web of all possible knowledge, Geddes tried to open up a space for the practical development of a social project leading to the spiritual ascent of man.'[57] This description by Michiel Dehaene of Geddes's efforts fits the masque projects well, and such leading to spiritual ascent is very clear in the parts of the masque that refer to India. Geddes's interest in properly representing Indian culture is reflected in the fact that by the time he was developing the masques (and following the untimely death of Sister Nivedita) he clearly thought of Ananda Coomaraswamy as an advisor.

COOMARASWAMY, INDIAN ART AND CELTICISM

It is possible that Geddes first met Coomaraswamy through Nivedita, but the more likely contact is C. R. Ashbee.[58] Geddes had known Ashbee since the early years of the Arts and Crafts movement, and in 1908 visited him at Broad Campden in Gloucestershire.[59] Coomaraswamy worked in close proximity with Ashbee, and what may well be the first mention of Coomaraswamy in *The Studio* is in a 1907 article by Ashbee on the conversion into a residence of the Norman Chapel at Broad Campden. This project was carried out for Coomaraswamy and his English wife, Ethel Partridge, who was herself a talented weaver. Ashbee notes the appropriateness of Coomaraswamy's presence: 'his collection of Sinhalese arts and crafts, upon a history of which he is at present engaged, is curiously fitted to the character of the building in which it is placed'.[60] The history that Ashbee mentions, *Mediaeval Sinhalese Art*, was published in 1908. It was not only Coomaraswamy's first major art historical statement, it was also printed on the very press used by William Morris for the *Kelmscott Chaucer*, a fact that Coomaraswamy is at pains to note in the preliminary pages of the book.[61] A further point of interest is that the drawings of the Norman Chapel for the 1907 *Studio* article were the work of Ashbee's assistant, Philip Mairet. On the break-up of

Coomaraswamy's marriage, Mairet was to marry Ethel, but it is Mairet's role as another generalist of the period with direct links to Geddes, as well as to Coomaraswamy and Ashbee, that I would note here.[62] Mairet was to remain in touch with Geddes for the rest of his life; indeed, in the 1920s he contributed illustrations to Geddes's *Makers of Future* series.[63] In the 1950s Mairet's book *Pioneer of Sociology: The Life and Letters of Patrick Geddes* made a valuable contribution to the understanding of Geddes.[64] Furthermore, he was a translator of Mircea Eliade and Jean-Paul Sartre, and wrote on the psychology of Alfred Adler.[65] Sister Nivedita was also present at Broad Campden, and Coomaraswamy's biographer Roger Lipsey draws attention to an invitation card from 1908 which asked friends to meet Nivedita at the Coomaraswamy house (i.e. the Norman Chapel), where she would give a talk on 'The Life of Indian Women in Relation to Religion, Education, and Nationalism'.[66]

There is a strong current of awareness of Celtic culture in Coomaraswamy's work. An unexpected indicator of this is to be found in his work as a geologist, for in one of his earliest publications, from 1903, one finds him reflecting upon the composition of the marble to found in the Inner Hebridean islands of Tiree and Iona. It is hard to imagine Coomaraswamy visiting these islands without taking an interest in their history and folklore. One can therefore date the beginnings of his informed awareness of Celtic cultures from this period. I noted earlier the distinction between the literary Anglo-Celticism of William Sharp and the scholarly but engaged collection of Gaelic oral material by Alexander Carmichael.[67] It is no less than one would expect of Coomaraswamy that he shows little interest in Sharp, but is consistently interested in the work of Carmichael and that of Carmichael's predecessor, John Francis Campbell of Islay.[68] His references to Carmichael and Campbell are usually in the form of comparative observations and footnotes relating to the oral tradition, and they continue throughout his life. An early mention of Carmichael is in Coomaraswamy's seminal cultural nationalist text *Essays in National Idealism* of 1909, a book in which he also quotes a letter from Geddes at length.[69] But as late as 1947, the year of his death, one finds him referring to Campbell in a note to his review essay 'Of Hares and Dreams'.[70] In 1945 he quotes not only Campbell and Carmichael but also J. G. Mackay and W. J. Watson.[71] Watson held the chair of Celtic at the University of Edinburgh from 1914 to 1938, and Coomaraswamy's reference is to introductory material for the 1940 publication of *More West Highland Tales*, a further book of Gaelic material collected by Campbell, translated into English by Mackay. This demonstrates that Coomaraswamy not only valued such dual-language Gaelic and English texts but also kept abreast of what was being published. A further indication of his informed awareness is found with respect to Celtic art. In early 1947 one finds him in correspondence with the outstanding analyst of Celtic patterns, George Bain, writing from Boston to compliment him on his 'excellent booklets on Celtic Art'.[72] These booklets focused on designs found on Pictish stone carvings of the eighth to tenth centuries. They were in due course gathered together and expanded to take in related manuscripts made by Gaelic speakers, such as *The Book of Kells*. Bain's canonical book, *The Methods of Construction of Celtic Art*, was published in 1951. He laments Coomaraswamy's death in his preface.[73]

Thus Coomaraswamy's interest in Celtic cultures can be traced throughout his career. However, his references to Geddes at the time of the Masques of Learning emphasise the latter's role not with respect to Celtic cultures but as an educator of global significance whose principles of action were relevant in an Indian context. In his essay 'Education in Ceylon', Coomaraswamy writes that he would like to see Ceylonese students gaining experience in a variety of European countries and studying Indian history and culture, but 'above all I should like them to come under the personal influence of men like Professor Geddes and women like Sister Nivedita'.[74] The paper was given in 1911 and collected in *Art and Swadeshi*, published in Madras in 1912. These cultural-nationalist essays are more or less continuous with those found in *Essays in National Idealism* of 1909, and it is interesting to find Geddes figuring in both collections, reinforcing the view that he was firmly part of an Indian intellectual milieu long before he set foot in India. Also mentioned in 'Education in Ceylon' is the distinguished educational thinker Michael Sadler (senior), who was a contributor to *The Sociological Review* along with Geddes, Coomaraswamy and Nivedita.[75]

In mid-March 1912 Geddes sent Coomaraswamy a copy of *The Masque of Learning*. By the end of the same month, in a letter of 30 March, Coomaraswamy had written an enthusiastic reply.[76] He requested extra copies of the book and also suggested a few changes to the Indian section; indeed, it is clear from this letter that Geddes's adoption of the term 'guru', in the more extensive text published as *The Masque of Ancient Learning* later in 1912, was due to Coomaraswamy's advice. An interesting use of the word, bringing into conjunction the experience of East and West, is Geddes's description of Plato:

We return then to our Masque, with its retrospect, its confrontation of the essentials of Indian and of Occidental thought. For now, when we come to our Western origins, we shall find them more Oriental then we knew. Aristotle, although our foremost system-maker, was never a curriculum-enforcer. His master, Plato, was no mere professor, tutor or don; but a Guru. Above all, Socrates, though, beyond all men, remembered in history as the questioner, was never an examiner: he lived and laboured as a birth-helper of the spirit, not a chooser of spirits slain.[77]

In his letter of March 1912, Coomaraswamy also notes Geddes's intention to repeat the masque in London and suggests that any Indian scene could be entrusted to the India Society.[78] Earlier that year Coomaraswamy had written to Geddes hoping that if he was visiting London he would be able to attend lectures held by the society.[79] The significance of the India Society should be noted here. It had come into being as a direct result of the cultural struggle between those, like Sir George Birdwood and Vincent Smith, who argued that India had no indigenous tradition of fine art and that what seemed to be so was merely the result of Hellenic borrowings, and those, like Coomaraswamy, Nivedita and E. B. Havell, who defended Indian art as an art in its own right since prehistoric times. Consciously or not, the Birdwood-Smith position was a way of justifying the status quo of political and cultural domination of India by Europe, by denying to Indian culture an identifying high-status feature of European culture, namely a firmly rooted tradition of fine art. As Partha Mitter notes, 'what Birdwood failed to see was the patronising element in his admiration for Indian art'.[80] In 1910 the issue had come to a head after a lecture by Havell at the Royal Society of Arts in London, and the controversy generated led to the foundation of the India Society. In 1913 Coomaraswamy published his major statement on the matter, 'Indian Images with many Arms', in *The Burlington Magazine*, and it was to be reprinted in his key essay collection *The Dance of Shiva*.[81] Geddes absorbed the arguments of Havell and Coomaraswamy into his own thought, and this can be seen in his lecture 'The Divine Ideal in Indian Art'.[82] It only exists in typescript and is undated. However, it extends one's appreciation of Geddes's awareness of pan-Asian issues, for here he also makes reference to the travels in Tibet of the Japanese Buddhist monk Ekai Kawaguchi, whose book *Three Years in Tibet* had been published by the Theosophical Publishing House in Adyar in 1909 or 1910.[83] In its title Geddes's paper is close to Havell's book *The Ideals of Indian Art*, published in 1911; indeed, that book contains a chapter entitled 'The Development of the Divine Ideal'.

In *The Masque of Ancient Learning* Geddes not only draws on Coomaraswamy's advice, but he makes explicit reference to

Coomaraswamy, writing of Indian art as a subject 'which we have long failed even to recognise, much less to penetrate or comprehend'. He continues: 'the artist with Abanindranath Tagore and Mrs. Herringham, the teacher with Mr. Havell, and the critic with Sister Nivedita and Dr. Coomaraswamy, are at length revealing to us its beauty and its significance'.[84] The list is typical of Geddes, for he begins with practice, proceeds though education and concludes with commentary. Christiana Herringham is a significant inclusion. She was both expert in the materials of painting and the leader of a group of artists, mostly Abanindranath Tagore's students, who had copied the Ajanta frescoes in 1910. The making of the Ajanta copies was a crucial aspect of the reappraisal of the non-Hellenic origins of Indian art.[85]

ENCYCLOPAEDIAS AND COLLEGES

Of particular interest from the perspective of this book are Geddes's emphases in the masques on academic foundations and encyclopaedias; he is at pains to emphasise the Scottish contribution in an international context. For example, he pointedly incorporates the history of the Scots College in Paris, whose revival he had promoted since 1890.[86] Writing of the time of Robert Bruce, when Scotland was re-establishing itself in the wake of resistance to English invasion, Geddes notes that:

> the independence of Scotland had to be maintained – and this intellectually and spiritually, as well as by the material forces of the Franco-Scottish Alliance which was thus necessitated. Hence Bruce founded, or rather renewed, for the Scots at Paris the famous Collège des Ecossais. This, as rebuilt from its medieval ruin by the Bishop of Moray in the sixteenth century, is preserved in good repair by the French Commission of National Monuments.[87]

It is clear that Geddes is interested here in educational policy-making as much as in historical reflection for he continues that it well rewards a visit, but he emphasises that while the historical side – mediaeval, renaissance or Jacobite – is of interest, crucial also should be the recovery of the Scots College by the Scottish universities: 'Oxford will retain its manifold advantages and charms; but our renewal of contacts with the continent generally, with France and with Paris . . . remains none the less the paramount desideratum of our Scottish Universities.'[88]

He brings the narrative up to date by invoking the importance of France for major nineteenth-century scientific pioneers active in Scotland:

> Nor is this less obvious in recent than in ancient times. Loyal to Glasgow and to Cambridge as he was, Kelvin was wont to tell how he was aroused to his

life-work during his wander-year in Paris. In the same way we have heard Lister speak, and with even more obvious indebtedness, of his 'honoured master, Louis Pasteur.' Nor do our Scottish painters forget their debt to Barbizon. Analogous impulses are still available, as every wandering student knows.[89]

Geddes the generalist deftly weaves art and science together into his vision of Scotland and France, and he goes on to note that the beginnings of a new collegiate residence in Paris were linked to developments in Edinburgh in the late 1880s (i.e. University Hall). Earlier in the masque, Geddes had dramatised the thirteenth-century growth of the University of Paris in an imaginary debate between the Dominican Thomas Aquinas and the Franciscan Duns Scotus, bringing into the picture not only Dante, as a student, but precursors of Aquinas and Duns Scotus such as Alcuin and that pan-European interdisciplinary thinker *par excellence* Michael Scot.[90] The last mentioned was, it will be recalled, a key figure in the Ramsay Lodge Murals by John Duncan painted over fifteen years earlier. Geddes underlines the Franco-Scottish connection by noting that 'no less than seventeen of the Rectors of the University of Paris in these early centuries were Scots'.[91] He continues by making the whole argument current again, emphasising that these historical notes should guide contemporary policy:

> As its students came from far and near, so its teachers wandered, finding welcome and audience in the schools of the towns they settled in. Such 'university extension centres,' as we should now call them, became the natural germ of new permanent universities, as at Orleans, Oxford, and so on; just as again in our time has happened at Dundee, and in various English cities.[92]

Geddes would pick up on that theme of the wandering scholar, important both to his own education and to his way of teaching others, in his farewell lecture to University College Dundee, an exploration of which forms my concluding chapter. These snippets from Geddes's masque texts indicate his pedagogical purpose in no uncertain terms. Similarly, when he turns to the Enlightenment: 'In this Encyclopedic and Philosophic age Scotland is once more in Europe, as of old, one of the Great Powers, of Culture.'[93] But as always, Geddes is using the past to look to the present and the future: 'Of other individuals of this group much might be said', but the one he singles out is 'Ferguson with his *Study of Civil Society*, as contributing not a little to lay the foundations of sociology – a work in which Kames, Monboddo and others were also participating'.[94] Geddes not only indicates his admiration for Adam Ferguson as a precursor of sociology, but takes for granted a direction of research into the Scottish Enlightenment which wasn't fully recognised until 1966, with the publication of Duncan Forbes's edition of Ferguson's *An Essay on the History of Civil Society*.[95] And, generalist that he is, Geddes continues, 'all this illustrious company was, on its own frank confession, outshone by the brilliant passage of Robert Burns'. He notes Burns as both thinker and man of action, and that these aspects were 'potentially even greater than the poet' affording 'a conclusive expression of our ever-recurrent thesis and leit-motif, that of the rise from nature to thought, as from labour into song'.[96]

Interest in the masques was not limited to Edinburgh and London. In March 1913 the rector of Morgan Academy, Dundee wrote to Geddes in the following terms: 'I learn from the Dundee Advertiser that performances of your "Masque of Learning" are to be given in London next week – our holiday week. Would it be possible with your permission and advice for pupils – present and former – to reproduce it at the Morgan?'[97]

PUBLISHING AND THE MASQUES

The content of the masques was echoed in the Edinburgh publishing of the time. As we have seen, the early days of University Hall in Edinburgh had been accompanied by the development of 'Patrick Geddes and Colleagues' as a publisher, and the printed versions of the masques are both a revisiting and extension of the cultural purposes of that imprint. But in contrast to the earlier work the published versions of the masques were produced in paper covers at low cost and in high print runs.[98] The emphasis was on mass distribution rather than quality of production. Thus they have none of the substantial Arts and Crafts quality of *The Evergreen* or the books of the Celtic Library. However, Geddes's pioneering of high-quality books for a general readership in the 1890s was not forgotten in Edinburgh and at the very time that he was engaged in developing his masques, Thomas Noble Foulis and his brother Douglas were taking further Geddes's example of affordable books inspired by Arts and Crafts principles. An interesting transitional publication is *The Meditations of Marcus Aurelius Antoninus*, published neither by Geddes nor by Foulis but by Otto Schulze in Edinburgh in 1902. The printer of that edition was W. H. White, who had printed most of the Celtic Library titles published by Geddes and Colleagues. The translation is based on that issued in 1742 by the Foulis Press in Glasgow, a publisher that acted as an inspiration for the activities of the distantly related Foulis brothers in Edinburgh a century and a half later.

The translation was reworked for publication by William Robertson Smith's biographer, George Chrystal.[99]

From 1903 onwards, T. N. Foulis (as the publisher became known) had begun to experiment with what has been described as 'a curious mix of private press influences and mass production techniques'.[100] Crucial input came from the artist Joseph Simpson, whose involvement with Edinburgh publishing was noted in *The Studio* in 1905: 'he can touch nothing without bringing it beauty and distinction, whether it be the designing of the cover of a book or arranging its very type'.[101] Simpson had trained at Glasgow School of Art and his contacts in the city provided a pool of outstanding illustrators for T. N. Foulis publications. These included D. Y. Cameron, Kate Cameron, Annie French and Jessie M. King. From south of the border, a key illustrator was Frank Brangwyn. Assessing the contribution of T. N. Foulis in *The Thistle* in 1913, W. Keith Leask commented that 'Scottish publishing has too long lacked two essentials – Nationality and Artistic Distinction. We trust that the firm thus auspiciously inaugurated will remember this; equally sure that, if it does, Scotland will remember it.'[102] In 1912 T. N. Foulis published another Geddes project, *The Blue Blanket: An Edinburgh Civic Review*. Like *The Evergreen*, it was a four-part journal of substantial and diverse content. It took its title from the traditional symbol of the Edinburgh craft guilds.[103] The contributors included Marjorie Kennedy-Fraser, writing not only on Hebridean song but also on a proposal for a music school for Scotland. A. P. Laurie wrote on higher education in Edinburgh, while images came from Otto Schlapp, John Duncan, Gerald Smith (including a notable portrait of Geddes himself) and Bruce J. Home.[104] Marion I. Newbigin made scathing comments on the psychological inadequacies of 'the perfected factory'.[105] Frank Mears's contribution, 'Public Monuments', was jointly authored with the architect Ramsay Traquair, son of Phoebe Traquair.[106] Two other papers prefigure Mears's later activities. One, 'Huntly House in the Canongate', by Charles Guthrie, describes the building which Mears was to restore in 1927. The other, 'Zoos' by Arthur Thomson, puts the case for the Edinburgh Zoological Garden that Mears was to design in 1913, in collaboration with Geddes's landscape-architect daughter Norah and Geddes himself. Norah Geddes married Frank Mears in July 1915. Among the images reproduced in *The Blue Blanket* was John Duncan's drawing *Cuchulainn*, which was used in 1917 as the frontispiece of Marjory Kennedy-Fraser's second volume of *Songs of the Hebrides*. Another image by Duncan published in *The Blue Blanket* was *Riders of Sidhe*, a finished drawing of Duncan's painting of the same title which, as has been noted, links directly to the Celtic aspect of Geddes's masques.

It is possible that Geddes was the link between T. N. Foulis and Ananda Coomaraswamy, but it is equally likely that Coomaraswamy, as an advocate of small presses which shared Arts and Crafts production values, was well aware of Foulis anyway.[107] Whatever the case, in 1913 T. N. Foulis was the publisher of Ananda Coomaraswamy's *Arts and Crafts of India and Ceylon*.[108] This was very much in line with a strong Eastern current in the publisher's list that complemented the Scottish material. For example, one finds selections from the Persian Sufi poet Hafiz side by side with Dr John Brown's *Rab and his Friends*, and Fitzgerald's translation of *Omar Khayyam* juxtaposed with Dean Ramsay's *Reminiscences of Scottish Life and Character*. In the same series as Coomaraswamy's book were Flinders Petrie's *Arts and Crafts of Ancient Egypt* and Stewart Dick's *Arts and Crafts of Old Japan*. An illustrated edition of Kakuzo Okakura's *Book of Tea* would follow in 1919. One may speculate that it was through Coomaraswamy's American links that Foulis became the publisher of the first British edition of Kakuzo Okakura's book. Coomaraswamy began work at the Museum of Fine Arts in Boston in 1917. Okakura had worked there previously. The involvement of Geddes's circle in Foulis's publications continued into the 1920s; indeed, one the last significant Arts and Crafts publications by T. N. Foulis was *The Wind in the Pines* issued in December 1922. This was a selection of material mainly from *The Evergreen* published to raise money for the Outlook Tower. The wider scope of Coomaraswamy's thought is found in another Foulis project, *Essays in Post-Industrialism: A Symposium of Prophecy concerning the Future of Society*, but it seems never to have got further than page proofs due to the outbreak of the First World War.[109] Coomaraswamy's co-editor was A. J. Penty, the pioneer of guild socialism.[110] In his 1922 book *Post-Industrialism*, Penty credits Coomaraswamy with the invention of the term.[111]

A dimension to Foulis as a publisher which underlines its Arts and Crafts ethos was the commissioning of type designs from the French designer Georges Auriol, whose work the Foulis brothers had first encountered in the pages of *The Studio*. An early use of Auriol's lettering for a Foulis cover was for A. P. Laurie's *Materials of the Painter's Craft*, a book that in both its form and its content underlines Foulis's commitment to maintaining crafts skills among artists.[112] The author was the son of Simon Somerville Laurie, whose relevance to the Scottish intellectual tradition I have noted in chapter one. Where the father was professor of education at the University of Edinburgh and an authority on the generalist thinking of Comenius, the son became principal of Heriot-Watt College in Edinburgh and professor of chemistry at the Royal Academy in London. Like his father, A. P. Laurie was a distinguished educationist, not least in his role as editor of *The Teacher's Encyclopaedia*, which contains a substantial section on Comenius.[113] Laurie's *Materials of the Painter's Craft* would certainly have been of interest to John Duncan, who was experimenting with tempera at that time; indeed, Laurie's later standard work, *The Painters Methods and Materials* published in 1926, acknowledges the help of both John Duncan and Phoebe Traquair.

A specifically Celtic revival aspect of the Foulis list emerges with the publication of the Iona Books series from 1912 onwards. These books had a simple handmade paper wrapper printed with an interlace cross design, presumably by Joseph Simpson. As with Geddes's *Evergreen* in the 1890s, Scots language played a significant part here, not least in the first of the Iona Books, Annie H. Small's *A Scottish Anthology*. The next year Fiona Macleod's essay on Iona, *The Isle of Dreams*, was reprinted as one of the Iona Books, and another of the series was the Gaelic *Comasan Na H-Urnuigh* by the Rev. Malcolm Macleod. This was a dual-language text, translating another Iona book, *The Possibility of Prayer*, and it was intended as a teaching aid for those who wanted to learn Gaelic. It contained an illustration by John Duncan. Other Iona books from that year were G. E. Troup's *Saint Columba* and James Wilkie's *Saint Bride*. Another title published that year was *Irishmen All* with illustrations by Jack B. Yeats, the key Irish artist of his generation and brother of W. B. Yeats.[114]

T. N. Foulis's publications mirror, to a significant degree, Geddes's concerns in his masques to make art and ideas properly available on a historically informed everyday level. Introducing his final masque publication, Geddes noted: 'for history is no mere retrospect of the past, nor excavation in it: what it reveals to us, above all, is the past still working on within our apparent present'.[115]

13

Looking East

BAHA'ISM AND THEOSOPHY

THE PERIOD DURING WHICH Geddes was organising his masques for Edinburgh and London also sheds light on the breadth of his spiritual interests through the visit to Edinburgh of the leader of the Baha'i faith, Abdul Baha, which took place in January 1913. Abdul Baha's key Edinburgh contact was Jane Whyte, who had visited him in 1906 while he was still under house arrest in Acre.[1] By 1913 Jane's husband, Alexander, was principal of New College in Edinburgh, the divinity school of the United Free Church. Like Jane, he was keenly interested in the ideas of religious unity that underpinned the Baha'i faith. Alexander Whyte's own teachings were inclined to both mysticism and ecumenicism, and he welcomed the Baha'i leader as an honoured guest at his manse in Charlotte Square.[2] Geddes was closely involved. Abdul Baha visited the Outlook Tower and later spoke to public meetings elsewhere in Edinburgh, the first of which was held at the Freemasons' Hall under the auspices of the Esperanto Society. It was chaired by Geddes's supporter John Kelman, who was also assistant minister to Alexander Whyte at St George's West Church in Edinburgh.[3] A subsequent meeting was chaired at New College by Geddes himself.[4] After Abdul Baha's visit, meetings to explore Baha'ism further were held at the Outlook Tower. They were led by the Baha'i Alice Buckton. Buckton's presence adds a further dimension to an understanding of Geddes's networks, for she is best known as a key figure in the revival of Glastonbury as a spiritual centre (see below).[5] In 1922 Geddes would ask his son-in-law Frank Mears to design a Baha'i temple for Allahabad in India. The temple was never built, but the project indicates Geddes's sustained interest in Baha'i ideals. Abdul Baha's third public meeting in Edinburgh was held at the premises of the Theosophical Society. The society had provided a key introduction to Abdul Baha's work for Scottish readers in a substantial article about Baha'ism that had appeared in the October 1911 issue of *Theosophy in Scotland*, and Theosophical ideas were another key aspect of Geddes's spiritual make-up.

There is no evidence that Geddes himself was a Theosophist, even though many of his symbolic devices such as *Lapis Philosophorum* share much with Theosophical imagery. But he certainly had a number of connections to Theosophy, and the non-doctrinaire religious generalism that underlies a Theosophical approach is something that would have attracted Geddes. In addition, he would have been well aware that the woman to whom he taught biology in London during his student days, Annie Besant, had become president of the Theosophical Society in 1907. But a close connection to Theosophy at the time of Abdul Baha's visit was through John Duncan. Duncan had joined the Theosophical Society in October 1909.[6] Many of his works have a strong Theosophical dimension, while remaining firmly within the purview of Geddes's Celtic revival. Indeed, Duncan blends the two currents seamlessly in *The Riders of the Sidhe* from 1911, a work that as we have seen was significant for Geddes's masques. In that painting Duncan's interest in the complementarity of East and West is explicit in his use of a symbol common to both traditions, the swastika, on one of the bridles.[7] The adoption of that symbol by the Nazi party in Germany in the 1920s and the atrocities perpetrated in that party's name make it hard to think of the swastika in a European setting without our view being distorted, yet almost a decade after Duncan painted *Riders of the Sidhe* it was still possible to begin a popular essay: 'A good many people have never heard of the Swastika. It is an emblem or device such as is the Cross or the Crescent.'[8] An immediate context for Duncan in 1911 would have been the presence of the swastika on the seal of the Theosophical Society.

Such Theosophical links deepen an understanding of Patrick Geddes's wider milieu. In April 1912 Duncan married a fellow Theosophist, Christine Allen, and both transferred to the newly formed Orpheus (Arts) Lodge in Edinburgh in October 1912.[9] His new wife was English, a member of the Celtic revival network which had been developing around Glastonbury. There had been connections between the Edinburgh and Glastonbury groups in the person of William Sharp from the early years of the century. After Sharp's death in 1905, his influence was still felt in Glastonbury, in particular through his writings (in the person of Fiona Macleod) about the Celtic goddess-come-saint, Bride. An example is *The Coming of Bride*, a play from 1914 by Alice Buckton who, as noted above, had led classes on Baha'ism the previous year at the Outlook Tower in Edinburgh after Abdul Baha's visit. Music was also significant: an operatic version of Fiona Macleod's drama *The Immortal Hour* by Rutland Boughton was first performed in Glastonbury in the summer of 1914, and it became very popular.[10]

The intriguing story of the Glastonbury Celtic revival has been told by Patrick Benham in *The Avalonians*.[11] A key moment was the concealment in 1897 of a glass bowl reputed to have belonged to Christ. The place of concealment was St Bride's Well at Glastonbury. As Benham explains, that action by Dr John Goodchild (a friend and correspondent of William Sharp in both his guises) set off a train of events which gave new energy and direction to the interpretation of Glastonbury as a spiritual centre. At the heart of that interpretation was the figure of Bride, both in her Christian and pre-Christian forms.[12] In 1906 the bowl was found by two sisters, Janet and Christine Allen. Six years later, Christine was to marry John Duncan.[13] Bride had been a presence for Duncan in the 1908 pageant in Edinburgh in which so many of Geddes's family and wider circle took part, but from the time of his marriage onwards his commitment to Bride became strong, notably in his *St Bride* from 1913, which is now in the National Gallery of Scotland. It is interesting to note that a version of that painting has recently come to light in America. It is linked to Duncan's enduring friendship with the educator Ellen Gates Starr of Hull House in Chicago.[14] Starr had been well enough known to Geddes in 1901 for him to send her his best wishes in a letter to Jane Addams.[15] The enduring links between Geddes's circle and that of Addams deserve more attention, and Duncan's role seems to be pivotal.

In a work from 1917, *The Coming of Bride*, Duncan shows Bride in her pre-Christian form as the bringer of Spring. By contrast, the 1913 work shows a version of the story that the young Bride was transported by angels from Iona to be present at the birth of Christ. This version had been the subject of a piece by Fiona Macleod published in *The Evergreen* in 1895.[16] It reflects the synthesis of pre-Christian and Christian belief that was at the heart of the early Celtic church. The mixing of traditions was underlined in another work by Duncan, *Christ Walking on the Sea*, which has as a border inscription a quote from St Columba: 'My Druid is Christ'.[17]

In the years following his joining of the Theosophical Society in 1909, Duncan produced a number of his most important symbolist works. After the faerie world of *Riders of the Sidhe* from 1911 came a scene from Arthurian legend, *Tristan and Isolde*, painted in 1912. *St Bride* was painted in 1913, *The Adoration of the Magi* in 1915 and *The Coming of Bride* in 1917. Duncan's versions of Bride as saint and goddess are encapsulated in Anne Ross's comment that 'Bride was ultimately a pagan Goddess who became a Christian saint, as was so often the case.'[18] There was substantial interest in this phase of Duncan's work in *The Studio*. Thus *Tristan and Isolde* was given a full-page colour illustration,[19] while both *The Coming of Bride*[20] and *The Riders of the Sidhe* were reproduced in full-page black-and-white plates.[21] As noted above, a preparatory drawing of *Riders of the Sidhe* had been printed in 1912 in Geddes's interdisciplinary magazine *The Blue Blanket*.[22]

Just as with Geddes, we can see all Duncan's work of this period in terms of the mutual illumination of Eastern and Celtic cultures. In one of his notebooks he writes of oriental art as having 'more for me than occidental' and of thinking that Celtic ornament expressed Buddhistic ideas.[23] His interest in this is at its clearest in *The Adoration of the Magi*, another work reproduced in *The Studio*.[24] The setting is a Hebridean shore and Duncan includes a Yin-Yang symbol on the robe of the elderly Mage, a very early use of that symbol in Western art. The Chinese symbol of the complementarity of active and passive forces became popular in the West in the 1960s and there are clear continuities of thought between Geddes's time and the later period. For example, Geddes inspired Lewis Mumford, who in turn inspired Theodore Roszak, who – in books such as *The Making of a Counter Culture*, published in 1969 – provided intellectual definition for the later period.[25] Further emphasising his breadth of spiritual view, later in his career John Duncan was to make a portfolio of teaching images of the life of the Buddha at the behest of a Theosophical foundation, the Ananda College in Colombo.[26]

MODERNISM AND REVIVAL

At the end of chapter nine I drew attention to Geddes's interdisciplinary cultural nationalism as the basis for his view of international co-operation, and the 'looking East' of both Geddes and Duncan in the period immediately before the First World War is a further example of that attitude. With respect to Duncan's Theosophy, it is worth recalling how many of the progressive artists of the time were linked to Theosophy. One was the Russian Nicolas Roerich, who would attract the attention of Amelia Defries in her book *Pioneers of Science*.[27] Others included the Finnish painter Akseli Gallen-Kallela (who had decorated the Finnish pavilion for the Paris exhibition of 1900) and his compatriot Pekka Halonen.[28] Another example is Duncan's younger contemporary the Dutch artist Piet Mondrian, who joined the Amsterdam branch of the Theosophical Society in 1909, the same year that Duncan joined in Edinburgh.

Such consideration helps to further illuminate the combination of traditionalism and modernism that one finds in Geddes. For example, in 1910 John Duncan visited his younger contemporary the modernist painter J. D. Fergusson in Paris. He found the experimentation of the Parisian art scene inspiring, writing in his notebook of two lessons to be learned 'from the New Art of Paris'. First of all, to paint in 'full colour glow force richness light' (that whole unpunctuated list capitalised for emphasis), and secondly, to 'trust oneself, to dare to be original. Do it the way you want it to be for yourself.'[29] Thus Duncan welcomed the Parisian explorations with open arms, whatever the contrast with his own work. These comments occur in a notebook in which Duncan explores his ideas for *Riders of the Sidhe* and, however surprising it may be, one can thus see that symbolist painting taking modernism as a guide. The friendship of John Duncan and J. D. Fergusson reminds us again that cultural revivalism and modernism are deeply interlinked; indeed, one can see them as two sides of the same early twentieth-century coin. Similarly for Geddes: on the one hand he was a thorough-going modernist, father of town planning and so on, yet on the other hand he was an equally thorough-going cultural revivalist. For Geddes there was no conflict, for a sustainable future required an understanding of the past. Thus his modernism did not simply learn from the past, it depended on it. Another Scot who held that attitude was Charles Rennie Mackintosh. One can recall again Geddes's comment about Mackintosh's Glasgow School of Art building: 'never was concrete more concrete, steel more steely', yet at the same time never was architecture more informed by history.[30]

J. D. Fergusson was later to emphasise that his own modernism had a Celtic dimension.[31] His clearest Celtic-focused works are his illustrations for *In Memoriam James Joyce*, written by Geddes's admirer the poet Hugh MacDiarmid.[32] A further Celtic revival resonance in Fergusson's circle is the fact that the woman who became his lifelong partner, the pioneer of modern dance Margaret Morris, created the Celtic Ballet company in Glasgow in 1940. She had performed at Glastonbury in 1913, the same year she met Fergusson in Paris.[33] Morris was also a close friend of Charles Rennie Mackintosh and Margaret Macdonald; indeed, Mackintosh designed a dance theatre for her in Chelsea in 1920. It was never built but an indication of their enduring friendship can be found in Morris's gift of her book on dance notation to Macdonald and Mackintosh in 1928.[34] In 1931 Morris was co-director along with Geddes of a summer school at Château d'Assas, a building acquired by Geddes to complement the activities of the Scots College he had established at Montpellier.[35]

CELTIC REVIVAL AND BENGALI REVIVAL

As noted in chapter nine, a key point in Geddes's intellectual journey to the East was his contact in Paris in 1900 with Swami Vivekananda, Sister Nivedita and Jagadis Chandra Bose. Nivedita and her circle remained an important part of Geddes's international milieu.[36] At the time of her death in 1911, Bengal-based Nivedita was writing a book which would be published as *Myths of the Hindus and Buddhists*. The task of finishing the book was undertaken by Ananda Coomaraswamy. It has an additional importance because its inclusion of images by Abanindranath Tagore and his students marks a wider appreciation of the art of the Bengal School of painters. The importance of that school and the contributions of Nivedita, Tagore and Coomaraswamy had been noted by Geddes in his *Masque of Ancient Learning*.[37] Nivedita had facilitated that work, just as Geddes had facilitated the work of Celtic revival artists in Scotland a decade earlier. *Myths of the Hindus and Buddhists* was published in London in 1913 by Harrap, and the growing popular interest in Hinduism and Buddhism during this period is clear from the fact that it had competition from Donald A. Mackenzie's *Indian Myth and Legend*. Mackenzie's book also made use of Bengal School illustrations, although in a minor role.[38] It was published by Gresham, a London imprint of the Scottish publisher Blackie, and in the same series Mackenzie was author of *Myths of Crete and Pre-Hellenic Europe*, in which the illustrations were supplied by none other than John Duncan. Echoing Duncan's sense that Buddhist and Celtic art were connected, Mackenzie was to make very clear his interest in the relationship between the two with the publication in 1928 of his speculative *Buddhism in Pre-Christian Britain*. It should also be noted that like J. D. Fergusson, Mackenzie connects to a later generation of cultural revivalist activity in Scotland. As a poet he makes an appearance in Hugh MacDiarmid's anthology *Northern Numbers*, published by T. N. Foulis in 1920. This work became a cornerstone of the post-Geddes cultural revival in the Scotland of the mid-twentieth century.

SCOTTISH INTERDISCIPLINARITY IN INDIAN REFLECTION

Thus, with the benefit of hindsight, Geddes's journey to India in 1914 at the age of sixty seems inevitable. Helen Meller has made the point that Geddes was influenced in his move by the prospect of extra income for his projects.[39] That is no doubt true, but from a cultural perspective Geddes's Indian journey could not have been more appropriate. It began to draw to an end his intensive involvement with Scotland, although one remarkable act remained, namely his parting lecture to University College Dundee, of which more in my concluding chapter.

Geddes's planning activities in India deserve more detailed attention than I can give them here, but from the Scottish cultural viewpoint of this book I note that in 1916 Geddes was on the point of making possible architecture for India by Charles Rennie Mackintosh. Gavin Stamp quotes from one of Mackintosh's letters, written from London in July 1916, in which he refers to 'playing around with Professor Geddes at his Summer Meeting here' and again in early August 'I have had a tentative offer from the Indian government to go out there for some six months . . . to do some work in reconstruction where they want me to do the architecture.'[40] Geddes's advocacy of Mackintosh thus resulted in an intriguing 'almost' of world architecture. It got as far as the drawing board and some of the drawings have been identified.[41] Alan Crawford has noted allusions to Indian architecture in Mackintosh's designs, finding direct reference to the gates of the Great Stupa at Sanchi in his drawing of an arcaded street front.[42] This opens up an interesting line of Arts and Crafts thinking, for Crawford points out that one of the gates had been illustrated in W. R. Lethaby's *Architecture, Mysticism and Myth*, published in 1891. The strong Scottish dimension of Lethaby's book should be noted. Lethaby quotes from James Macpherson's *Poems of Ossian* to illustrate a point about the motion of the sun,[43] and he turns frequently to analyses of the mythological and the cosmic by William Robertson Smith, James Frazer and, most often of all, Andrew Lang.[44] Thus Lethaby picked up on the strong current of mythological analysis among late nineteenth-century Scottish thinkers, a current central to the thinking of both Geddes and Mackintosh. Lethaby also draws on the thinking of the pioneer of Indian architectural history, James Fergusson, another Scot. Mackintosh's interest in Lethaby's book is well established, but one should also bear in mind his direct interest in the works of Fergusson, for these would have given him a more detailed source. While Lethaby provides an attractive silhouette, Fergusson illustrates and discusses the Sanchi gates in detail in the third volume of his *History of Architecture*, published in 1876 and republished in 1910.[45]

Although the invitation to Mackintosh bore no direct fruit, Geddes's work opened the door for architecture and design in India by Mackintosh's godson Eckart Muthesius. Muthesius was the son and student of Mackintosh's supporter Hermann Muthesius, and Francis and Jessie Newbery of Glasgow School of Art were his godparents.[46] As Geddes found it harder to get commissions in those parts of India under direct British rule, he turned increasingly to the 'rulers of the native Indian states to keep up his flow of commissions'.[47] One of these was the Maharaja of Indore, who commissioned Geddes's report *Town Planning Towards City Development: A Report to the Durbar of Indore*, and it was at Indore in the 1930s that Muthesius was to find employment. Geddes completed his Indore report in 1918 and it can be set alongside the Dunfermline report of 1903 as among the most significant of all his publications. Like the earlier work, it has an appreciation of the history of ideas and the importance of education at its heart. Thus, alongside Geddes's rethink of the historic old town of Indore and planning for new suburbs, alongside observations on drainage and disease control, is a seventy-three-page section devoted to 'A New University for Central India'.[48]

The unrealised projects by Mackintosh can be seen in the context of what has been called the 'careful attention to Indian/Hindu tradition' of Patrick Geddes.[49] It is perhaps too easy to blame the disruption of the First World War for the fact that such projects were never realised for, as noted above, although Geddes made a contribution to the planning of British-controlled India he was not known for seeing eye to eye with the Imperial establishment.[50] Nevertheless, the war must have been a significant factor. The war also led to heartbreak for Geddes when his eldest son and talented assistant Alasdair was killed on the Western Front. Alasdair's ability to interpret and sketch landforms had distinguished him as a balloon observer in the Royal Flying Corps. He died near Arras in April 1917. This personal tragedy was compounded for Geddes by the illness and death of the major support of his life, his wife Anna, then engaged in organising Geddes's first Indian summer meeting at Darjeeling.[51] Anna died in June. So in three months in 1917 Geddes lost, in difficult circumstances, two of the people closest to him. Despite these events, his subsequent achievements were extraordinary.

However difficult the Darjeeling summer school must have been for Geddes, it was there that he began to cooperate actively with Rabindranath Tagore.[52] Tagore and Geddes shared a great deal. Both took interdisciplinarity for granted, both were extraordinarily active in cultural revival, both appreciated the importance of ecology. And in due course Arthur Geddes would work closely with both men. Whatever the tragedy of Alasdair's death, Geddes was fortunate in his surviving children. In his eldest child Norah and her younger sibling Arthur he had two children who, like Alasdair, understood the significance of their father's work and, crucially, were interested in it to a professional standard. Norah became a landscape architect and Arthur became a geographer. It is to Norah that we owe key biographical insights into her father.[53] It is to Arthur that we owe the preservation of much of Geddes's archive. From the point of view of interdisciplinary thinking, Arthur's work with Rabindranath Tagore is of particular interest. As the links between his father and Tagore developed, Arthur acted as a conduit of information between them. Later, as a geographer at the University of Edinburgh, he made a significant academic contribution to the understanding of the regional geography of India and then applied his skills to Scotland, in particular to the Highlands and the Western Isles.[54] He was also, like his father, a committed cultural revivalist. To that end he translated some of Tagore's songs into English, beginning – with Tagore's blessing – in the 1920s with five songs from Tagore's drama *The Dark Chamber*.[55] These works and others were performed at the Edinburgh Festival in 1961.[56] From a Scottish perspective, Arthur can thus be seen as a link between the Celtic revival of the 1890s and the folk revival of the 1960s. One of Arthur's students at Edinburgh University, Walter Stephen, recalls a meeting of the Edinburgh University Geography Society in the early 1960s: 'About a hundred students were assembled when the speaker failed to turn up. Unperturbed, Arthur Geddes handed round copies of his own collection of Scottish Songs, *Songs of Craig and Ben*, tuned up his fiddle and led this mass of students in an hour of exuberant music and singing.'[57] The second volume of *Songs of Craig and Ben*, a selection of Gaelic songs that Arthur had translated with a view to performance, was published in 1961.[58] It contains an introduction which explores the songs of the Highlanders of Scotland in terms of the valley section. Not only that but there is substantial reference to Alexander Carmichael's *Carmina Gadelica*, and to the same author's *Deirdire*, bringing the revival of the 1960s into explicit correspondence with that of the years around 1900. With Arthur's help, Patrick Geddes went on in the years after the Darjeeling summer meeting of 1917 to help Tagore develop his college-communities at Santiniketan and Sriniketan. Bashabi Fraser's editing of the correspondence between Geddes and Tagore has put research into that meeting of minds onto a secure footing.[59]

SUNWISE TURN AND *THE DANCE OF SHIVA*

During this period the friendship between Patrick Geddes and Ananda Coomaraswamy continued. One can symbolise it in two letters from

Coomaraswamy to Geddes. The first dates from 1914 and in it he notes that Geddes will be surprised by the address from which he is writing, for it is Geddes's own flat in Ramsay Garden in Edinburgh. The letter was sent to Geddes when he had himself left for India.[60] The second letter was written three years later and in it one finds Coomaraswamy travelling in New Mexico and giving Geddes an enthusiastic account of encountering Hopi and Navaho cultures in the form of art and crafts, architecture and dances, which he thinks would all interest Geddes immensely.[61] He continues his letter by noting, almost as an aside, his appointment as keeper of Indian art at the Museum of Fine Arts in Boston. It was to be the key academic appointment of his life. The institution had previously employed Kakuzo Okakura, and Boston lay at the heart of a New England Transcendentalist hinterland sympathetic to cultural revival. Coomaraswamy had quoted Thoreau as early as 1909 when he invoked him, along with William Morris, in the preface of his early cultural nationalist statement, *Essays in National Idealism*. In this New England context, Coomaraswamy would influence the mythologist Joseph Campbell; indeed, Campbell chooses Coomaraswamy – along with James Frazer, Max Muller, Durkheim, Jung and the Church – as one of his six definers of the function of mythology.[62]

This American east coast mythological pluralism is linked in an interesting way to Geddes's milieu in Scotland. At first sight Ananda Coomaraswamy's *Dance of Shiva*, published in New York in 1918, a book still regarded as a definitive restatement of Indian culture, might seem an unlikely place to find a Celtic revival dimension. When the book is considered as a text alone, there is none. However, when the publisher is taken into account a different story emerges. The publisher of *The Dance of Shiva* was the radical New York arts bookshop Sunwise Turn. To anyone aware of Celtic tradition, the phrase 'sunwise turn' is immediately familiar. It transpires that the name of the bookshop has its origin in the visit of the American ethnomusicologist Amy Murray to the folklore collector Father Allan McDonald of Eriskay in 1905. John Duncan had been encouraging Marjory Kennedy-Fraser to come to Eriskay to transcribe Gaelic song since he had himself first visited the island in 1904. Duncan's friend Bessie Macarthur recounts that it was Duncan discovering Amy Murray beginning to collect material from Father Allan that spurred him to encourage Kennedy-Fraser to come out without delay, 'or the songs would be heard in America before they were heard in Scotland'.[63] In the event, Father Allan died later in 1905, but not before Amy Murray had collected and learned

to sing a significant amount of material under his guidance. In his biographical essay on Father Allan, John Lorne Campbell credits Murray
with a 'greater ability in taking down the old Gaelic airs than any other
transcriber'.[64] Her book, *Father Allan's Island*, published in America
in 1920 and in Scotland in 1936 is a valuable account, not least
because of its recording of Father Allan's view of Fiona Macleod. In
1905 Macleod was still not commonly identified as William Sharp and
Murray quotes McDonald as saying 'she's got all her Gaelic – where it
isn't wrong – out of Mrs Mary Mackellar's *Guide to the Highlands*'.[65]
That is fair comment. Murray herself showed a real commitment to
Gaelic, reproducing a number of songs including the beautiful *Ailein
Duinn*. She also shows a sensitivity to Scots (as distinct from Gaelic)
language, criticising friends in the Borders for restricting use of it in
their children.[66] Furthermore, she recognised Gaelic song as an aspect
of what we would now call world music, citing comparisons ranging
from friends of Vivekananda being reminded of his chanting in Sanskrit
to analogies being made with singing from the Kentucky mountains.[67]
In due course Alan Lomax and Paul Strand would retrace Murray's
footsteps in the Western Isles.

An account of the naming of the bookshop comes from one of its
founders, Madge Jenison, who recalls that 'Amy Murray had drawn up
in the net of her Gallic [sic] wisdom the name *The Sunwise Turn*. "They
do everything deasal (sunwise) here" – Father Allen [sic] had told her of
the people of Eriskay – "for they believe that to follow the course of the
sun is propitious. The sunwise turn is the lucky one." '[68] Thus, one of
the most influential early twentieth-century texts on Indian culture, *The
Dance of Shiva*, was published by a New York bookshop named after a
Celtic ritual observed in the Outer Hebrides of Scotland. Coomaraswamy
no doubt enjoyed the spiritual, cosmic and folkloric resonance between
the name of his publisher and the title of his book. Further insight
into Coomaraswamy and the bookshop can be found in Allan Antliff's
Anarchist Modernism. Antliff notes the significance of the Sunwise Turn
bookshop as a meeting place for modernist thinkers in New York, and
he makes clear that it was the centre of Coomaraswamy's intellectual
and artistic life in the city.[69] It was a remarkable place. Madge Jenison
notes that the young Peggy Guggenheim 'in a moleskin coat to her heels'
was an assistant in the shop,[70] and that Thorsten Veblen 'always bought
interesting things'.[71] The year after the publication of Coomaraswamy's
The Dance of Shiva, Sunwise Turn published the first English translation
of Rainer Maria Rilke's book on Rodin, complete with a frontispiece
photograph of the sculptor by Gertude Kasebier.[72]

Jenison notes a direct link between her remarkable bookshop and Patrick Geddes. When he left Dundee to take up his chair of sociology and civics at the University of Bombay, he entrusted the provision of the American titles for his departmental library to Sunwise Turn.[73]

14

A Farewell Lecture

MOVING ON FROM DUNDEE

I CONCLUDE THIS BOOK with a consideration of Geddes's farewell
lecture to University College Dundee given in the midsummer of 1919.
It is one of the most telling statements of his commitment to a generalist
approach to education. At the same time it marks his physical moving
on from Scotland.[1] By that date Geddes's most distinguished students
and assistants had taken up leadership roles elsewhere in the developing
disciplines of geography and ecology, and in 1917 his old friend D'Arcy
Thompson had left Dundee for St Andrews. Thus Geddes's lecture can
be seen as marking the conclusion of a remarkable period of over three
decades of biological research at University College Dundee.[2] We owe its
transcription and publication to Amelia Defries, who devotes an entire
chapter to it in her book on Geddes.[3] It remains one the most concise
yet comprehensive statements of his thinking, a measured statement of
the need for interdisciplinarity in the interests of both intellectual and
ecological survival. Geddes displays time after time the generalist ethos
that informed his activities throughout his career. The lecture allows
one to appreciate George Davie's description of Geddes as teaching with
the 'old philosophical bias' characteristic of Scottish education.[4] And it
is nothing if not an example of Davie's other description of Geddes as
'forward looking'.[5]

Geddes takes as his title *Biology and Its Social Bearings: How a
Botanist Looks at the World*. He sets the academic context not in terms
of the lecture theatre but in terms of travel: 'we botanists learn to see
the great world, and try to make each student his own traveller, gaining
his own widening vision of the world'. From those wandering Dundee
scholars had come 'the best maps, as yet, of Lowlands and Highlands',

and those maps had led on to the best maps for Britain as a whole. In these introductory remarks Geddes takes care to note the role of the Ecological Society and emphasises A. J. Herbertson's regional work at Oxford and Marcel Hardy's *Atlas of World Vegetation*.[6] Herbertson had been Geddes's assistant in Dundee in 1892, as had Hardy from 1900 to 1905, and, as noted in chapter eleven, one of the key early figures in the British Ecological Society was Geddes's student William Smith. Thus Geddes begins by underlining his own research leadership at University College Dundee and at the same time demonstrates the development from Dundee of wider academic networks. He then goes to the heart of the actual practice of ecology and the awareness that it requires:

> How many people think twice about a leaf? Yet the leaf is the chief product and phenomenon of Life: this is a green world, with animals comparatively few and small, and all dependent upon the leaves. By leaves we live.[7]

Geddes reminds us that it is folly to consider money as more important than nature, noting that 'the world is mainly a vast leaf-colony . . . and we live not by the jingling of our coins, but by the fullness of our harvests'.[8] Thus life, not money, is the issue. An obvious enough point, one might think, but one that Geddes knew required stating and restating. The phrase 'by leaves we live' encapsulates Geddes's ecological message. He did not himself employ it as a motto, but it has taken on a life of its own.

The wide-ranging aspect of the lecture will already be clear. Among much else, Geddes goes on to reflect on his little practical success in transforming Dundee through the introduction of gardens, with the exception of his own teaching garden within the precincts of the university and those of 'a school here and there'.[9] As noted in chapter eleven, it would have been a different story had his 1906 plan for a botanic garden linking to a network of other gardens leading into the heart of the university campus gone ahead. He gives his vision of planning as in essence a process of the growing of gardens: 'It grows on from small gardens to semi-public ones . . . and thence to parks and boulevards, and so to better houses for all upon their course or beyond it.' And he continues: 'that is, indeed, the way in which planning has actually grown: even the magnificent circles and avenues of Paris are but the outcomes of clearings through the forest'. But he notes wearily the tendency to confuse this organic notion of town planning 'with the destructive (or at best, mechanical) activities of engineers'.[10]

When Defries is not directly transcribing Geddes's words, she interjects illuminating descriptions of his actions as he took up plant after plant from a substantial temporary garden set up around his lectern.[11] Early on in the lecture Geddes pulled from it an acanthus leaf, the stock in trade of the architectural sculptor over millennia: 'here in the garden around us there is a whole world of beauty for designers to choose from – only the edge of which has been touched'. A little later he adds an intriguing comment: 'perhaps we are afraid to let an artist into a university: he might do something!'[12] The presence of the temporary garden is clear throughout the lecture as Geddes demonstrates point after point from nature: 'see here, how tall and strong this [lily] is growing, seeming to be using all its energies for itself'. And then contrasting it with other lilies in different phases of growth: 'this one in bloom is now living for its species – flowering magnificently', and he continues: 'its "purity" is the very opposite of the sexless misunderstandings [of the lily in] the past. It is the fullest splendour and frankness of sex in nature, naked and not ashamed.'[13] That transfer of the notion of purity from repressed religion to fecund nature is given philosophical context some pages later when he comments: 'what is Bergson's *Élan Vital* but an appreciation of how flowers grow?', continuing: 'our older theories were more of how artificial flowers got put together, or how anglers' flies were dressed: mechanically beautiful, no doubt, but not real live flowers or flies!'[14] He also invokes Bergson's notion of duration rather than clock time to make sense of the time needed for cultural growth: 'ideas . . . take time to grow, and while a sower knows when his corn will ripen, the sowing of ideas is, as yet, a far less certain affair'.[15]

Geddes takes that sowing of ideas as his cue to discuss education at all levels, beginning with what is for him the start of all education, namely wonder: 'star-wonder, stone and spark wonder, life-wonder, folk-wonder . . . the stuff of astronomy and physics, of biology and the social sciences'. Having linked that fundamental impulse to its disciplinary framework, he shifts to method via 'the fundamental place of Nature Study, and of our Surveys'. He then makes the move straight back to the aesthetic as the basis for all disciplinary study: 'to appreciate sunset and sunrise, moon and stars, the wonders of the winds, clouds and rain, the beauty of woods and fields – here are the beginnings of natural sciences'.[16]

It is hardly surprising to find Geddes scathing when he notes the manner in which impulses to education are undermined in both children and adults. Instead of overcontrolled teaching, for Geddes 'each child needs its own plot in the school-garden, and its own bench in the workshop; but it should also go on wider and wider excursions, and these increasingly of its own choice'.[17] And in the next sentence Geddes turns from the practical interdisciplinarity of garden and workbench and excursion to the theoretical interdisciplinarity of the academy, reflecting, in the passage I drew attention to in chapter one, on the interdependence of arts and sciences and how each should inform the other: 'we need to give everyone the outlook of the artist, who begins with the art of seeing – and then in time we shall follow him into the seeing of art, even the creating of it'. That crucial link between art and science is thus for Geddes the art of seeing which lies at the heart of initiating the scholar and the student, 'as in our Edinburgh Tower', into the outlook of astronomer and geographer, of mathematician and mechanic, of physicist and chemist, of geologist and mineralogist, of botanist and zoologist, 'and thence more generally, of the biologist'. And he continues, 'next, too, the anthropologist . . . and thus, too, the economist'.[18] Then he draws in history, philosophy and poetry, noting that 'this general and educational point of view must be brought to bear on every specialism'.[19] His generalist approach to education could hardly be more explicit. He sums up by returning to educational method and emphasising that the teacher's outlook should include all viewpoints and that those viewpoints must be at the service of the student: 'individual realization of these, and choice of life-interests and of occupation thus go to form the personal outlook of the student'.[20]

A little later he does not pull his punches with respect to the state of education as he perceives it in 1919: 'All this is in contrast, no doubt, to the current Nature-starvation of school and college, with their verbalism and empaperment.' And having noted one sort of educational starvation, Geddes proceeds to note that 'this also means moral starvation', which he attributes not least to the so-called 'best education', which leads to 'that paralysis of "good form," the real meaning whereof is the utmost possible inhibition', which, scathing once more, he calls 'the delight, so often, of the British parent, pedagogue, snob and fool!'[21] Once again his words seem as relevant now as they were then. He continues by noting that 'the mistaken and perverted order is still prevalent' and 'beginnings like Madame Montessori's, or our own at the Outlook Tower – of course with its complemental Inlook – are still far too few'. A remarkable appreciation of that 'inlook' aspect of the Outlook Tower had been written two years earlier by the poet Wilfred Owen, at that time a patient at Craiglockhart Hospital in Edinburgh.[22] For Geddes, as for Montessori, the recognition of the role of the senses is fundamental: 'The eye is predominantly important for this intellectual life (*Do you see?*) and the ear for emotional appeal . . . and as senses are thus deeply related in life, so with our ideas, our whole personality and powers'. And he continues that 'we must cease to think merely in terms of separated departments and faculties, and must corelate [sic] these in the living mind; in the social life as well – indeed, this above all'.[23] This advocacy of knowledge as emerging from a comparative process at once sensory, cognitive and social is fundamental to the Scottish generalist tradition as described by George Davie: 'according to which science and philosophy, individual selves, and our senses, each, at their own levels, reciprocally illuminate one another'.[24] Geddes continues: 'Intellectual education involves general and sensory, imaginative and artistic education: Re-education, Re-creation, and thus Reconstruction and the conception of Culture in its literal sense, of "to cultivate".'[25]

He concludes his lecture thus: 'And so – with art inspiring industry, and developing the sciences accordingly – beyond the attractive yet dangerous apples of the separate sciences, the Tree of Life thus comes into view.'[26] Remember that Geddes was speaking to an audience immediately after the First World War; indeed, his lecture was given before full demobilisation had taken place, as is clear from Defries's description, which refers to the 'many sturdy, khaki-clad figures' in the audience.[27] Thus his reference to 'the dangerous apples of the separate sciences' points the finger directly at the context-free approach to science that gives rise to societies driven by what would in due course be called the military-industrial complex. Again, his words seem even more to the point today, a century after he spoke them. But he leaves us with the Tree of Life coming into view, an echo of his *Arbor Saeculorum* from *The Evergreen*, an image of potential integration of thinking, action and spiritual purpose with which to consider the past and look to the future. Soon he would be urging the cultural reunification of Europe through 'the education of travel'.[28]

AN ITALIAN CODA

The power of Patrick Geddes's final lecture to University College Dundee lies not only in his advocacy of the interdependence of the local and the international, and his sense of responsibility to the ecological wellbeing of the planet – it is also a trenchant reminder of the enduring importance of his Scottish interdisciplinary view. One gets an indication of that significance from the outstanding Italian architect and urban planner Giancarlo De Carlo. Speaking in Edinburgh seventy-five years after Geddes's Dundee lecture, De Carlo underlined the value of Geddes's interdisciplinary approach as follows:

Here in Scotland, in Scottish culture, from what I have read and I have studied, I think you have one educational pillar which is very important. It is what you call generalism ... I think we should go back to the idea of the general view, and in Scotland you have a good grounding in this approach, not least because of the work of Patrick Geddes.[29]

The purpose of this book is to enable thinking about Patrick Geddes from that generalist perspective. Whether considered through the lens of Jan Amos Comenius in the seventeenth century, Giancarlo de Carlo in the twentieth century or of Geddes's compatriots David Masson, Simon Somerville Laurie and James Clerk Maxwell in the nineteenth century, that approach continues to be relevant.

Notes

CHAPTER 1   Geddes and the Scottish Intellectual Tradition

1. Tagore, Rabindranath, foreword to Defries, Amelia, *The Interpreter: Geddes, the Man and his Gospel* (London: Routledge, 1927), p. xiii.
2. Davie, George, *The Democratic Intellect: Scotland and her Universities in the Nineteenth Century* (Edinburgh: Edinburgh University Press, 1961); Davie, George, *The Crisis of the Democratic Intellect: The Problem of Generalism and Specialisation in Twentieth-Century Scotland* (Edinburgh: Polygon, 1986).
3. Davie, *Democratic Intellect*, p. 13.
4. Ibid., p. 24.
5. Ibid., p. 22.
6. Founded in Edinburgh in 1768 and in Geddes's time still published in that city.
7. For exploration of Geddes from this perspective, see (i) Macdonald, Murdo, 'Patrick Geddes e l'intelletto democratico/Art and the context in Patrick Geddes's work', *Spazio e Società*, October/December 1994, pp. 28–45; (ii) Macdonald, Murdo, 'Patrick Geddes and Scottish Generalism', in *The City After Patrick Geddes*, edited by Volker Welter (Bern: Peter Lang, 2000), pp. 55–70; (iii) Macdonald, Murdo, 'Patrick Geddes and the Tradition of Scottish Generalism', *Journal of Scottish Thought*, Vol. 5, 'Patrick Geddes', 2012, pp. 73–88. For further context, see Craig, Cairns, 'George Davie and Scottish Idealism', *Journal of Scottish Thought*, Vol. 5, 'Patrick Geddes', 2012, pp. 13.
8. Masson, David, *The Life of John Milton*, Vol. 3 (London: Macmillan, 1873), pp. 213–14.
9. Laurie, Simon Somerville, *John Amos Comenius, Bishop of The Moravians, his Life and Educational Works* (London: Kegan Paul, Trench, Trubner & Co., 1881). For a recent overview of Laurie, see Templeton, Ian, *Simon*

Somerville Laurie: His Educational Thought and Contribution to Scottish Education. 1855–1909 (PhD thesis, University of Edinburgh, 2010).

10. Cf. Templeton, *Laurie*, p. 43: 'in a practical sense it seems likely that many of [Laurie's] ideas can be traced to a greater or lesser extent to Comenius'.

11. Laurie, *Comenius*, p. 217.

12. Geddes, Patrick, *Cities in Evolution* (London: Williams and Norgate, 1915), pp. 368–9.

13. See chapter eight.

14. Burnet, John, 'Humanism in Education', read before the Classical Association of Scotland, 30 May 1914, and collected in Burnet's *Essays And Addresses* (London: Chatto and Windus, 1929). The quotation is from p. 125.

15. Burnet, John, *Higher Education and the War* (London: Macmillan, 1917), p. 227.

16. Robson, B. T., 'Geography and Social Science: The Role of Patrick Geddes', *Geography, Ideology and Social Concern*, edited by D. R. Stoddart (Oxford: Blackwell, 1981), pp. 186–205, p. 200 and p. 203.

17. Ibid., p. 200.

18. Branford, Victor V., *Interpretations and Forecasts* (London: Duckworth, 1914).

19. See Scott, John and Bromley, Ray, *Envisioning Sociology: Victor Branford, Patrick Geddes and the Quest for Social Reconstruction* (Albany: State University of New York, 2013). For a concise version: Scott, John and Bromley, Ray, 'The Geddes Circle in Sociology', *Journal of Scottish Thought*, Vol. 5, 'Patrick Geddes', 2012, pp.121–34. For insight into the rejection of Geddes's generalist view of sociology as it became professionalised, see Law, Alex, 'Scholastic ambivalence and Patrick Geddes: a sociology of failed sociology', *Journal of Scottish Thought*, Vol. 5, 'Patrick Geddes', 2012, pp. 41–71.

20. Defries, Amelia, *The Interpreter: Geddes, the Man and his Gospel* (London: Routledge, 1927).

21. My thanks to George Davie, who first drew Robertson's work to my attention.

22. Robertson, Stewart Alan, *A Moray Loon* (Edinburgh: Moray Press, 1933).

23. Duncan Macmillan makes the point well in his essay 'A single-minded polymath – Patrick Geddes and the spatial form of social thought', *Edinburgh Review*, No. 88, 1992, pp. 78–88, especially pp. 81 and 87. See also Mike Small's 'City Planning and the Anarchist Tradition', *Anarchist Studies*, Vol. 6, No. 1, March 1998, pp. 64–9, especially p. 68.

24. Bateson, Gregory, *Mind and Nature: A Necessary Unity* (New York: Dutton, 1979), p. 8. Also note Waddington, C. H., *Behind Appearance: A study of the relations between painting and the natural sciences in this century* (Edinburgh: Edinburgh University Press, 1969), p. 243, 'man is

Argus with innumerable eyes, all yielding their overlapping insights into his one being, that struggles to accept them in all their variety and richness'.

25. Davie, *Democratic Intellect*, pp. 127–8.
26. Boardman, Philip, *The Worlds of Patrick Geddes: Biologist, Town Planner, Re-educator, Peace-warrior* (London: Routledge & Kegan Paul, 1978), p. 20. Boardman goes on to note that Geddes's teacher at Perth Academy, Rector Miller, would stand at the blackboard enthralled by a proposition of Euclid he had just completed and murmur: 'What beauty! What beauty!'.
27. For insight into Maxwell's thinking in relation to Davie's view of the Scottish geometrical tradition, see Paterson, Lindsay, 'Geometry and the Scottish Imagination', *Cencrastus*, No. 17, Summer 1984, pp. 30–7.
28. Macdonald, Murdo, 'Patrick Geddes – educator, ecologist, visual thinker', *Edinburgh Review*, No. 88, 1992, pp. 113–19; Macdonald, Murdo, 'Patrick Geddes: Visual Thinker', in *A Window on Europe: The Lothian in Europe Lectures 1992*, edited by Geraldine Prince (Edinburgh: Canongate, 1993), pp. 229–55; Macdonald, Murdo, 'The Visual Thinker: Patrick Geddes', *Edinburgh-Yamaguchi '95 Papers* (Yamaguchi: Yamaguchi Institute of Contemporary Art, 1996), pp. 50–60.
29. Stone, Norman, *Europe Transformed* (London: Fontana, 1983), p. 411.
30. An influential Free Church thinker of Geddes's own generation was Henry Drummond, much of whose work, for example his book *The Ascent of Man*, published in 1894, was devoted to expounding the theory of biological evolution in terms of Christian faith.
31. Kelman, John, *The Interpreter's House: The Ideals embodied in the Outlook Tower, Edinburgh* (Edinburgh: Oliphant, Anderson and Ferrier, 1905).
32. Defries, *Interpreter*, p. 186.
33. Ibid., p. 175.

CHAPTER 2   Geography, History and Place

1. Boardman, Philip, *The Worlds of Patrick Geddes* (London: Routledge, 1978), p. 277.
2. Ibid.
3. Robertson, Stewart Alan, *A Moray Loon* (Edinburgh: Moray Press, 1933), p. 4.
4. Coates, Henry, *A Perthshire Naturalist: Charles Macintosh of Inver* (London: T. Fisher Unwin, 1923). Introduction by J. Arthur Thomson and Patrick Geddes. Macintosh has the added distinction of being a key influence on Beatrix Potter's nature studies during her childhood summer holidays.

5. Geddes, Patrick, 'The Sixth of the Talks from the Outlook Tower: The Education of Two Boys', *The Survey*, Vol. 54, No. 11, September 1925, pp. 571–5 and 587–91. The reference is to p. 573. Geddes's 'Talks from the Outlook Tower' were published by *The Survey* in New York from February to September 1925. They are collected in *Patrick Geddes: Spokesman for Man and the Environment*, edited by Marshall Stalley (New Brunswick, NJ: Rutgers University Press, 1972), pp. 289–380.

6. Geddes, *Sixth Talk*, p. 573.

7. On his birth certificate Geddes's name is recorded as 'Peter'. That is a common Anglicisation of the Gaelic 'Pàdraig' i.e. 'Patrick'. My thanks to Finlay MacLeod of Shawbost for discussion. Walter Stephen has pointed out that Geddes used both 'Peter' and 'Patrick' as late as his teens. See Stephen, Walter, 'Where was Peter Geddes born?', in *A Vigorous Institution*, edited by Walter Stephen (Edinburgh: Luath, 2007), pp. 85–96, p. 87.

8. Geddes, Norah, *Memoir of Patrick Geddes*, NLS MS 10508, f. 163.

9. He finally retired in 1878 with the rank of captain in the Royal Perth Militia. Information courtesy of Thomas B. Smyth, Archivist of the Black Watch Museum, Balhousie Castle, Perth.

10. The regimental headquarters of the Black Watch is in Perth.

11. Macphersons, Camerons, Stewarts, MacGregors, MacIntyres, Grants, Robertsons, MacDonalds and Frasers were among those executed or transported. 'It is impossible to avoid a belief that the Hanoverian Government's vindictive punishment of the mutineers did much to encourage, fortify, and enlarge the menace, to itself, of the Jacobite rebellion of 1745': Linklater, Eric and Linklater, Andro, *The Black Watch: the History of the Royal Highland Regiment* (London: Barrie and Jenkins, 1977), p. 23.

12. MacRae, Donald, 'Adam Ferguson', in *The Founding Fathers of Social Science*, edited by Timothy Raison (Harmondsworth: Penguin, 1969), pp. 17–26, p. 26.

13. With respect to Ferguson's view, see the numerous references to him in *Ossian Revisited*, edited by Howard Gaskill (Edinburgh: Edinburgh University Press, 1991), for example pp. 14–15.

14. *The Poems of Ossian: Translated by James Macpherson, with Notes, and with an Introduction by William Sharp* (Edinburgh: Patrick Geddes and Colleagues, 1896).

15. Grant, Charles, *The Black Watch* (Oxford: Osprey, 1971), p. 24.

16. See, for example, Joseph Cundall's 1856 photograph of William Gardner, Donald McKenzie and George Glen (Scottish National Portrait Gallery).

17. Boardman, *Geddes*, p. 11. Other interesting examples can be found illustrated in Kenneth MacLeay's *Highlanders of Scotland*, a commission from Queen Victoria which was published in 1870. For an earlier and less

sporran-orientated view, see McIan, Robert Ronald and Logan, James, *The Clans of the Scottish Highlands* (London: Ackermann, 1845).

18. Grant, *Black Watch*, p. 39.
19. John Michael Wright's *Lord Mungo Murray* (Scottish National Portrait Gallery) dates from about 1680. See Macdonald, Murdo, *Scottish Art* (London: Thames and Hudson, 2005), pp. 46–8.
20. Geddes, Patrick, 'Notes for Lecture to Celtic Society', SUA T-GED 5/2/9, 22/10/97.
21. Geddes, *Sixth Talk*, p. 573.
22. Ibid.
23. Ibid.
24. Ibid.
25. Ibid.
26. As Stewart Robertson noted, he was later to say, 'You must make of your geography a geosophy.' Quoted by Robertson, *Moray Loon*, p. 3.
27. Geddes, Norah, *Memoir*, ff. 164 and 165.
28. It is interesting to note that 'Geddes's first ambition was to be an artist'. McGegan, Edward, 'Sir Patrick Geddes', *The Scottish Bookman*, Vol. 1, No. 4, December 1935, pp. 98–106, p. 100.
29. Ibid., f. 172. Geddes's interest in geology was expressed in his friendship with James Geikie, who would later succeed his brother Archibald as Murchison Professor of Geology at the University of Edinburgh. Geikie certainly inspired Geddes and may have taught him.
30. Buist, Robert C., 'The Newest Scots College: Some Account of the Work of Patrick Geddes', *The Scots Magazine*, Vol. 8, 1928, pp. 321–4, p. 322.
31. Boardman, *Geddes*, p. 28.
32. Frost, Mark, *The lost companions and John Ruskin's Guild of St George* (London: Anthem Press, 2014), pp. 120–1.
33. Boardman, *Geddes*, p. 36.
34. Stoddart, D. R., 'That Victorian Science: Huxley's Physiography and its Impact on Geography', *Transactions of the Institute of British Geographers*, No. 66, Nov. 1975, pp. 17–40, p. 23.
35. For an assessment of Spencer from a source close to Geddes, see Thomson, J. Arthur, *Herbert Spencer* (London: Dent, 1906).
36. Boardman, *Geddes*, p. 35; See also Meller, Helen, *Patrick Geddes: Social Evolutionist and City Planner* (London: Routledge, 1990), pp. 28–9.
37. Geddes, Patrick, 'Introduction' to Liveing, Susan, *A Nineteenth-Century Teacher: John Henry Bridges* (London: Kegan Paul, Trench, Trubner and Co., 1926), p. 1.
38. For Ingram's influence on Geddes's economic thinking, see Campbell, Gordon, 'Patrick Geddes – a theory of value', *Edinburgh Review*, No. 88, 1992, pp. 63–72, p. 66. Ingram would later be a member of the consultative committee for Geddes's 1896 summer meeting.
39. Meller, *Geddes*, p. 63.

40. For a characterisation of generalist thinking in terms of the illumination of such blind spots, see Macdonald, Murdo, 'The democratic intellect in context', *Edinburgh Review*, No. 90, 1993, pp. 59–60.
41. Lacaze-Duthiers had in 1858 made a zoological discovery which would have appealed to the mixture of science and religion in Geddes's background, namely that the Mediterranean mollusc *Murex trunculus* was the source of the ancient dye Tyrian Purple, which is referred to in the Bible.
42. For an account of Geddes's role in the Cowie project, see Slater, Anne-Michelle, 'The Light Before the Darkness', in *A Vigorous Institution: The Living Legacy of Patrick Geddes*, edited by Walter Stephen (Edinburgh: Luath, 2007), pp. 103–22.
43. Davie, George, *The Democratic Intellect* (Edinburgh: Edinburgh University Press), p. 151. For further context, see Davie, 'Victor Cousin and the Scottish Philosophers', in Davie, George, *A Passion for Ideas* (Edinburgh: Polygon, 1994), pp. 70–109.
44. Leslie on Ancient (i.e. geometrical) Analysis; see Davie, *Democratic Intellect*, p. 151. For more on Comte's translation (made for J. N. Hachette), see Pickering, Mary, *Auguste Comte: An Intellectual Biography*, Vol. 1 (Cambridge: Cambridge University Press, 1993), pp. 57–8. Comte disliked the project, referring to it as a 'bad book', but that may have been a reflection of his personal circumstances rather than a comment on Leslie's work. My thanks to Mary Pickering (personal communication) for discussing the point 'in looking at his letter to Valat . . . it may indeed have been "bad" in his opinion because it didn't bring him anything, such as money, fame, or connections'. For the context, see Pickering, p. 58 and n. 203.
45. Livingstone, David N., *The Geographical Tradition: Episodes in the History of a Contested Enterprise* (Oxford: Blackwell, 1993), p. 271. Livingstone draws on Desmond, Adrian, *The Politics of Evolution* (Chicago: University of Chicago Press, 1989). Note in particular Desmond's section: 'The French Morphologies and Their Scottish Importers', pp. 41–92.
46. In the early years of the twentieth century, Gérard de Lacaze-Duthiers, presumably a younger relation of Henri de Lacaze-Duthiers, was a significant anarchist thinker in a circle which included Cubists, Futurists and Neo-Symbolists, including the Scottish painter J. D. Fergusson. See Antliff, Mark, *Inventing Bergson: Cultural Politics and the Parisian Avant-garde* (Princeton: Princeton University Press, 1993), pp. 136–46.
47. Reynolds, Siân, 'Patrick Geddes's French Connections in Academic and Social Life: Networking from 1878 to the 1900s', in *Patrick Geddes: The French Connection*, edited by Frances Fowle and Belinda Thomson (Oxford: White Cockade, 2004), p. 71.
48. See Defries, Amelia, *The Interpreter: Geddes* (London: Routledge, 1927), pp. 123–44; Geddes, Patrick, *Cities in Evolution*, revised edition (London: Williams and Norgate, 1949), Appendix 1, pp. 194–213,

and for an expression of the ideas in development, Geddes, Patrick, *The World Within and the World Without: Sunday Talks with my Children* (Bournville: Saint George Press, 1905).

49. Boardman, *Geddes*, p. 50.
50. Geddes, Arthur, 'Geddes, Sir Patrick (1854–1932)', *Dictionary of National Biography*, compact edition, Vol. 2 (London: Oxford University Press, 1975), p. 2649.

CHAPTER 3   Arts, Crafts and Social Reform

1. Johnson, Jim, and Rosenburg, Lou, *Renewing Old Edinburgh: The Enduring Legacy of Patrick Geddes* (Glendaruel: Argyll Publishing, 2010).
2. *Edinburgh Social Union and Social and Sanitary Society, Twenty-eighth Annual Report*, November 1912, pp. 8–9.
3. Reclus, Élisée, *The Evolution of Cities* (Petersham, NSW: Jura Books, 1995), p. 67. Originally published in *The Contemporary Review*, 1895. For more on Geddes and Reclus, see Ferretti, Federico, *Anarchy and Geography: Reclus and Kropotkin in the UK* (Abingdon: Routledge, 2018), in particular chapter two, 'Editorial networks and the publics of science: building pluralist geographies'.
4. Geddes, Patrick, ed., *Viri Illustres* (Edinburgh: Pentland, 1884).
5. Geddes, Norah, *Memoir of Patrick Geddes*, NLS MS 10508, f. 186.
6. Haldane, Elizabeth S., *From One Century to Another* (London: Maclehose, 1937), p. 118.
7. Ibid., p. 112. See also Haldane's account of the founding of the Social Union, published as a memorial to Octavia Hill, *Social Union, Twenty-eighth Annual Report*, p. 4.
8. Geddes includes Ferrier in *Viri Illustres*.
9. Haldane, Elizabeth S., *The Scotland of Our Fathers: A Study of Scottish Life in the Nineteenth Century* (London: Maclehose, 1933), pp. 132–3.
10. *Edinburgh Social Union, First Annual Report*, 1885, p. 6. The annual reports 1885–95 are collected in a bound volume in the Edinburgh Room of Edinburgh Central Library: HV 250 E23 S. The minute book 1885–92 (which also contains press cuttings) is available at QY HV 250 E23S.
11. *Edinburgh Social Union, Second Annual Report*, 1887, pp. 9–10.
12. *Edinburgh Social Union, First Annual Report*, 1885, pp. 10–12.
13. Ibid., p. 10.
14. Geddes, Patrick, 'John Ruskin: Economist', in *The Round Table*, edited by H. B. Baildon (Edinburgh: William Brown, 1887).
15. *The Round Table* drew together work by members of the Symposium Club at Edinburgh University. See Mavor, James, *My Windows on the Street of the World*, Vol. 1 (London: Dent, 1923), p. 208.
16. While a student in London, Geddes had taught Bradlaugh's daughter

science subjects at that time not available to women. Another of Geddes's women students in London was Annie Besant.

17. Whyte, Alexander, *Newman: An Appreciation* (Edinburgh: Oliphant, Anderson and Ferrier, 1901).

18. Barbour, George Freeland, *The Life of Alexander Whyte* (London: Hodder and Stoughton, 1923), p. 238.

19. *Dante: Illustrations and Notes* (Edinburgh: Privately Printed, T. and A. Constable, 1890). Phoebe Traquair also designed the cover, frontispiece and title page image. The accompanying notes were by John Sutherland Black. Traquair also produced a number of decorations for covers of Whyte's books, including a portrait roundel of Cardinal Newman. Among other notable contributions by Traquair to religious books of mystical inclination is her frontispiece to Grace Warrack's edition of Julian of Norwich's *Revelations of Divine Love* (London: Methuen, 1901). Later she provided a number of works for Warrack's *From Isles of the West to Bethlehem* (Oxford: Blackwell, 1921).

20. Geddes, Patrick, *Every Man his own Art Critic – Manchester Exhibition, 1887* (Manchester: John Heywood, 1887); Geddes, Patrick, *Every Man his own Art Critic – Glasgow Exhibition, 1888* (Glasgow: William Brown, 1888).

21. Geddes, *Every Man – Glasgow*, p. 57.

22. In a seminal work on the nature and history of modernist design, Norman Potter notes Geddes as a 'synthesiser of ideas and a social activist, much concerned with user-participation'. Potter, Norman, *What is a Designer: Things. Places. Messages* (London: Hyphen Press, 1989), pp. 103–4. Potter recalls Ashbee's comments that Geddes's work 'has ever a touch of prophecy' and 'when all's said and done . . . his prophecy is likely to sound the farthest'.

23. Hall, Peter, *Cities of Tomorrow* (Oxford: Blackwell, 1988), pp. 137–48. Hall also notes Geddes's link to Lewis Mumford. However, he misunderstands the degree of communication between Mumford and Geddes, claiming that they corresponded little after their uneasy first meeting in 1923. For the true position, see Novak, Frank G., ed., *Lewis Mumford and Patrick Geddes: The Correspondence* (London: Routledge, 1995). For a concise introduction to Geddes in his wider anarchist context, see Ward, Colin, *Anarchism: A Very Short Introduction* (Oxford: Oxford University Press, 2004), p. 87.

24. Reilly, John P., *The Early Social Thought of Patrick Geddes*, PhD thesis (New York: Columbia University, 1972), p. v.

25. John Kelman would become one of the first published advocates of Geddes's educational activities at the Outlook Tower in Edinburgh. Kelman, John, *The Interpreter's House: The Ideals embodied in the Outlook Tower, Edinburgh* (Edinburgh: Oliphant, Anderson and Ferrier, 1905).

26. Jane Whyte would be a key link to Baha'ism: see chapter thirteen.
27. Reilly, *Geddes*, p. 219.
28. Ibid.
29. For the wider context of this appointment, see Swinney, Geoffrey, 'Wyville Thomson, *Challenger*, and the Edinburgh Museum of Science and Art', *The Scottish* Naturalist, Vol. 111, 1999, pp. 207–24.
30. Caw, James L., 'The Art Work of Mrs Traquair', *The Art Journal*, 1900, pp. 143–8, p. 144.
31. Cumming, Elizabeth, *Phoebe Anna Traquair 1852–1936* (Edinburgh: Scottish National Portrait Gallery, 1993), p. 9.
32. Ibid., p. 13.
33. *Edinburgh Social Union, Second Annual Report*, 1887, p. 10.
34. Brown, Gerard Baldwin, 'Some Recent Efforts in Mural Decoration', *Scottish Art Review*, Vol. 1, 1889, pp. 225–8, p. 228.
35. Ibid., p. 225.
36. Works by both artists were subjects of Geddes's writing in *The Scottish Art Review*. Geddes, Patrick, ' "The Wood Nymph" and "Silence": E. Burne Jones A.R.A., Dante Gabriel Rossetti', *The Scottish Art Review*, Vol. 1, 1889, pp. 155–6.
37. Reproduced in Cumming, *Traquair*, p. 15.
38. *History and Description of the Decorations by Mrs Traquair in the Mortuary of the Royal Hospital for Sick Children, Edinburgh*. Probably c. 1900. Reprint republished in 1993 by Department of Paediatric Pathology, Royal Hospital for Sick Children, Edinburgh. The misidentification of David Scott occurs on p. 9.
39. Those with Scottish and/or Celtic revival links represented in this wide-ranging collection include Hall Caine, Andrew Lang, Noel Paton, Marc-André Raffalovich, Ernest Rhys, William Bell Scott and James (B. V.) Thomson.
40. Pater, Walter, *Imaginary Portraits* (London: Macmillan, 1887). The panels are now in the National Gallery of Scotland.
41. Ruskin, John, *Praeterita*, Vol. 2 (Orpington: George Allen, 1887), p. 423. Quoted in Forrest, D. W., ed., *The Letters of John Brown, With Letters from Ruskin, Thackeray, and Others* (London: A. and C. Black. 1907), p. 287.
42. Cumming, *Traquair*, p. 17.
43. Crane, Walter, *An Artist's Reminiscences* (London: Methuen, 1907), p. 326. Crane noted Traquair's significance in the preface to *Of the Decorative Illustration of Books*, 3rd edition (London: Bell, 1904).
44. Traquair, Phoebe Anna, *Reproductions by me, of medallions painted by me, in a border on the walls of the song school, St. Mary's Cathedral, Edinburgh* (Edinburgh: Edinburgh University Library, 1897), Ms. Gen. 852. Also see Cumming, *Traquair*, cat. 43.
45. The Catholic Apostolic Church had its basis in the heresy of a minister

of the Church of Scotland, Edward Irving, sometime assistant to Thomas Chalmers. Irving died in 1832, and, notwithstanding his erstwhile differences with the Church of Scotland, is buried in the crypt of Glasgow Cathedral. See Strachan, Gordon, *The Pentecostal Theology of Edward Irving* (London: Darton, Longman and Todd, 1973).

46. *Edinburgh Social Union 10th Annual Report*, Nov. 1894, pp. 18–19.
47. Lockie, Katherine F., *Picturesque Edinburgh* (Edinburgh: Lockie, 1899), p. 127 and p. 175.
48. For Geddes's influence on Newbery, see Eadie, William, *Movements of Modernity: The Case of Glasgow and Art Nouveau* (London: Routledge, 1990), pp. 92–3.
49. McKay is remembered today for his book *The Scottish School of Painting* (London: Duckworth, 1906).
50. Not long afterwards, the architect Robert Lorimer returned to Edinburgh and in 1892 he became a member of the School of Artistic Handicraft, a Social Union grouping convened by Geddes's older colleague James Cossar Ewart, the professor of natural history at the University of Edinburgh. Baldwin Brown, Capper and Traquair were also members of the school.
51. This remarkable purpose-built building was threatened with closure in the early 1990s, but the decision was reversed thanks to a public campaign.
52. Smailes, Helen, *A Portrait Gallery for Scotland* (Edinburgh: Scottish National Portrait Gallery, 1985), pp. 67–8.
53. Cumming, *Traquair*, p. 24.
54. Hole, Elizabeth, *Memories of William Hole R.S.A. by his wife* (Edinburgh: Chambers, 1920), p. 85.
55. Ibid., pp. 75–6.
56. Cumming, *Traquair*, p. 17.
57. MacCarthy, Fiona, *William Morris – A Life of Our Time* (London: Faber and Faber, 1994), p. 505.
58. For more on the close relationship between Morris and Scheu, see frequent references in Thompson, E. P., *William Morris: Romantic to Revolutionary* (London: Merlin Press, 1976).
59. Reilly, *Geddes*, p. 201. Reilly's source is a typescript by the anarchist T. H. Bell, entitled 'Patrick Geddes', in the Joseph Ishill Collection in the Houghton Library of Harvard University. Reilly notes that it was catalogued under two different accession numbers: b MS Am 1614(234) and b MS 1614.1(303).
60. Oliphant, James, ed., *The Claims of Labour* (Edinburgh: Co-operative Printing Co., 1886). Along with Geddes and Morris, the line-up of lecturers included Alfred Russell Wallace.
61. Mavor, James, *My Windows on the Street of the World*, Vol. 1 (London: Dent, 1923), p. 199.
62. Ibid., p. 214.

63. *Scottish Art Review*, Vol. 1, No. 1. Quoted in Billcliffe, Roger, *The Glasgow Boys: The Glasgow School of Painting, 1875–1895* (London: John Murray, 1985), p. 228.

64. *Scottish Art Review*, Vol. 1, No. 11. Prior to this no editor is named, although see Billcliffe, *Glasgow Boys*, pp. 227–31. Vol. 2 starts in June 1889. Numbering of parts is consecutive with Vol. 1, that is to say Vol. 2 starts at No. 13.

65. Billcliffe, *Glasgow Boys*, pp. 227–31.

66. Mavor, *My Windows*, Vol. 1, p. 234. See also Craig, Cairns, 'Introduction: The Circles of Patrick Geddes – Patrick Geddes and James Mavor', *Journal of Scottish Thought*, No. 5, 'Patrick Geddes', 2012, pp. 1–15.

67. Indication of Mavor's editorial reach can be seen in a leaflet preserved in the National Library of Scotland (NLS APS.2.90.104), entitled *Scottish Art Review* 'announcements for 1889'. It lists papers by, among others, Baildon, Blackie, Blanc (Edinburgh-based Beaux-Arts architect and Social Union member), Baldwin Brown, Edward Carpenter, Walter Crane, Havelock Ellis, Patrick Geddes, Mavor himself, William Morris, James Oliphant (Social Union), Ernest Rhys, John M. Robertson, William Sharp, R. A. M. Stevenson, Mortimer Wheeler (father of the archaeologist of the same name) and Gleeson White. Images announced included works from George Clausen, Joseph Crawhall, James Guthrie, George Henry, John Lavery, Harrington Mann, Arthur Melville, W. G. Burn Murdoch, James Paterson, Alexander Roche and E. A. Walton.

68. Mavor, *My Windows*, Vol. 1, p. 235.

69. Cumming, Elizabeth, 'A note on Phoebe Traquair and an Edinburgh Dante', *Edinburgh Review*, No. 88, 1992, pp. 143–9.

70. This was also the period that Robert Louis Stevenson's cousin and close friend, R. A. M. Stevenson, established himself as a pioneer of modern art criticism with his study of Velázquez, published in 1895. For a biographical study of Stevenson and the impact of his criticism, see Denys Sutton's introduction to the 1962 edition.

71. Some sense of Blaikie's interdisciplinary abilities can be gleaned from a collection of obituaries, *Walter Biggar Blaikie (1847–1928)* (Edinburgh: Privately Printed, 1929). One of the pieces was by Geddes.

72. *Scottish Art Review*, Vol. 2, pp. 158–9.

73. Ibid., p. 174.

74. *Scottish Art Review*, Vol. 1, p. 210. The other papers so praised were by William Holman Hunt, Walter Crane, William Morris, Alfred Gilbert and T. J. Cobden-Sanderson.

75. Crane, Walter, *An Artist's Reminiscences* (London: Methuen, 1907), p. 326. The tower to which he refers is presumably James Court, for the Outlook Tower was not taken over by Geddes until 1892, three years after Crane's visit.

76. Ibid., p. 327; see also Glasier, J. Bruce, *William Morris and the Early Days*

of the Socialist Movement (London: Longmans, Green & Co., 1921), especially chapter ten, 'Edinburgh Art Congress and After'.

77. Cumming in Bowe, Nicola Gordon and Cumming, Elizabeth, *The Arts and Crafts Movements in Dublin and Edinburgh* (Dublin: Irish Universities Press, 1998), p. 17.

78. The *Transactions* of the congress, from which most of my information is derived, is yet another production of T. and A. Constable: *Transactions of the National Association for the Advancement of Art and its Application to Industry, Edinburgh Meeting* (London: 22 Albemarle Street, 1890).

CHAPTER 4   Early Years of University College Dundee

1. Boardman, Philip, *The Worlds of Patrick Geddes* (London: Routledge, 1978), p. 55.

2. Philip Abrams notes that White's 'interest in sociology was largely formed by his early friendship with Geddes and Branford'. Abrams, Philip, *The Origins of British Sociology* (Chicago: University of Chicago Press, 1968), p. 103.

3. Boardman, *Geddes*, p. 78.

4. *Dundee Naturalists' Society, Ninth Annual Report, 1881–82*. Dundee Public Libraries, Lamb Collection, DPL LC, 142(25), 4.

5. Davie, George, *The Democratic Intellect* (Edinburgh: Edinburgh University Press, 1961), p. 337.

6. The membership list published with the Dundee Naturalists' Society annual report for 1882–3 shows Geddes to be one of only seven honorary members, another of whom was James Geikie.

7. Kitchen, Paddy, *A Most Unsettling Person: An Introduction to the Ideas and Life of Patrick Geddes* (London: Gollancz, 1975), p. 84.

8. Ibid., p. 85.

9. Southgate, Donald, *University Education in Dundee* (Edinburgh: Edinburgh University Press, 1982), p. 31.

10. Kitchen, *Geddes*, p. 85.

11. Barbour, George Freeland, *The Life of Alexander Whyte* (London: Hodder and Stoughton, 1923), p. 203. Barbour is himself quoting from the biography of Robertson Smith by George Chrystal and James Sutherland Black. Barbour was the brother of Alexander Whyte's wife, Jane.

12. Ibid., pp. 203–4.

13. Ibid., p. 201.

14. Robertson Smith continued to be a figure of international reputation. Some indication of his overall influence can be found in the work of his younger contemporary Sigmund Freud, who devoted a substantial section of *Totem and Taboo* to a discussion of Smith's work (and to that of Smith's follower, James Frazer).

15. Kitchen, *Geddes*, p. 85.

16. Ibid., p. 86.
17. Nash, David, *Secularism, Art and Freedom* (Leicester: Leicester University Press, 1992), pp. 22–3.
18. Lutyens, Mary, *Krishnamurti: The Years of Awakening* (London: John Murray, 1975), pp. 13–14.
19. Mears, Norah G., *Intimations and Avowals* (Edinburgh: The Moray Press, 1944), p. vii.
20. Boardman, *Geddes*, p. 437, 437n.
21. 11 March 1885. *Dundee Naturalists' Society Twelfth Annual Report, 1884–5*, Dundee Public Libraries, Lamb Collection, DPL LC, 142(27).
22. Ibid.
23. The same year also saw Thompson elected to the Royal Society of Edinburgh. Both Geddes and Young were proposers. See *Former Fellows of the Royal Society of Edinburgh, 1783–2002*, Part 2 (Edinburgh: Royal Society of Edinburgh), p. 920.
24. Kitchen, *Geddes*, pp. 87–8.
25. NLS MS 10524, ff. 71,72.
26. Ibid.
27. Ibid.
28. For further context, see Southgate, *Dundee*. Geddes is noted as both 'locally-known' and as a 'beneficiary of the old Dundee YMCA science classes', p. 53.
29. Jarron, Matthew, 'Forget the silly notion that I'm here to teach you botany – Patrick Geddes at University College Dundee', in *The Artist and the Thinker: John Duncan and Patrick Geddes in Dundee*, edited by Matthew Jarron (Dundee: University of Dundee Museum Services), pp. 30–41, p. 30.
30. Meller, Helen, 'Understanding the European City around 1900: The Contribution of Patrick Geddes', in *The City after Patrick Geddes*, edited by Volker Welter and James Lawson (Bern: Peter Lang, 2000), pp. 35–54, p. 40.
31. Geddes, Patrick and J. Arthur Thomson, *The Evolution of Sex* (London: Walter Scott, 1899), p. 314.
32. Robertson, Stewart A., 'Patrick Geddes', in *A Moray Loon* (Edinburgh: Moray Press, 1933), p. 1. Also note that in 'The Moral Evolution of Sex', published in *The Evergreen* in 1896, Geddes and Thomson extended their argument to society and education, reflecting on the dynamic balance of competition and co-operation.
33. Baillie, Myra, 'The Grey Lady: Mary Lily Walker of Dundee', in *Victorian Dundee: Image and Realities*, edited by Louise Miskell, Christopher A. Whatley and Bob Harris (East Linton: Tuckwell, 2000), pp. 122–34, p. 124.
34. For example, NLS MS 10530, f. 209, from White to Geddes dated 20 September 1898, details the arrangement for a loan which seems to relate

to the completion of Ramsay Garden, and NLS MS 10533, ff. 222–4, a letter from White dated 12 November 1902, seems to relate to repayments of the same loan.

CHAPTER 5   Education, Anarchism and Celtic Revival

1. Geddes rented the tower from 1892 and acquired it in 1896. See Wallace, Veronica, 'Maria Obscura', *Edinburgh Review*, 88, 1992, pp. 101–9, p. 101. Wallace's paper focuses on the intriguing pre-Geddes history of the tower, as Short's Observatory.
2. Bowe, Nicola Gordon and Cumming, Elizabeth, *The Arts and Crafts Movements in Dublin and Edinburgh* (Dublin: Irish Universities Press, 1998), p. 24.
3. Boardman, Philip, *The Worlds of Patrick Geddes* (London: Routledge, 1978), p. 73.
4. Kelman, John, *The Interpreter's House: The Ideals embodied in the Outlook Tower, Edinburgh* (Edinburgh: Oliphant, Anderson and Ferrier, 1905).
5. *Edinburgh Social Union, 22nd Annual Report*, November 1906.
6. *University Hall, Edinburgh* (Edinburgh: Town and Gown Association, 1903), p. 3. The authorship of this document is unclear. It was prepared by Whitson and Methuen C. A., Secretaries, The Town and Gown Association, Limited. Geddes's supporter Thomas Whitson may well be the author.
7. Salmond, J. B., *Recording Scotland* (Edinburgh: Oliver and Boyd, 1952), p. 30.
8. NLS MS 10615 f1.
9. *University Hall*, p. 10.
10. Cf. 'Geddes was emphatically not a conservationist, but a passionate moderniser. As his own interventions in the Edinburgh Old Town showed, he would happily demolish or alter old buildings at will if they stood in the way of his wider cultural vision of the future': Glendinning, Miles and Page, David, *Clone City: Crisis and Renewal in Contemporary Scottish Architecture* (Edinburgh: Polygon, 1999), p. 35.
11. For an account of the building and ethos of Ramsay Garden, see (i) Johnston, Margo, 'Ramsay Garden, Edinburgh', *Journal of the Architectural Heritage Society of Scotland*, No. 16, 1989, pp. 3–19. Johnston also notes Mitchell's partner George Wilson as having an involvement; (ii) Leonard, Sofia G. and Mackenzie, J. M., *Ramsay Garden* (Edinburgh: Patrick Geddes Centre for Planning Studies, 1989).
12. David Macdonald drew my attention to the reports of Geddes in the minutes of the council of the Cockburn Association, and also to Peggy Bawden's 1982 note of Geddes's link to the association, in the association's archives.

13. Aitken had an involvement in a number of Geddes projects, notably as architect for his 1904 Dunfermline Report for the Carnegie Trust.

14. Whitson, Thomas B., 'Lady Stair's House', *Book of the Old Edinburgh Club*, Vol. 3, 1910, pp. 243–52.

15. Johnson, Jim, and Rosenburg, Lou, *Renewing Old Edinburgh: The Enduring Legacy of Patrick Geddes* (Glendaruel: Argyll Publishing, 2010).

16. NLS MS 10508A ff. 127, 7 November 1895.

17. Translated at my request, before I knew its source, by my Edinburgh University colleague Tom Schuller. The source is line 286 of Aeschylus's *Eumenides*.

18. For the radical context of this song, see *The Canongate Burns*, edited by Andrew Noble and Patrick Scott Hogg (Edinburgh: Canongate, 2001), pp. 512–16.

19. An appreciation of the contemporary importance of Élisée Reclus's ecological thought can be gleaned from *Liberty, Equality, Geography: The Social Thought of Élisée Reclus* (Oxford: Lexington Books, 1997), edited and translated by John Clark and Camille Martin. Geddes's view is put in the major obituary essay he wrote for *The Scottish Geographical Magazine*, Vol. 21, 1905, entitled 'A Great Geographer, Élisée Reclus (1830–1905)'. This appeared in two parts on pp. 490–6 and pp. 548–55. In 1901, in Vol. 17 of the same magazine, Reclus had published his paper 'The Teaching of Geography' (pp. 393–9), which includes reference to Geddes's Outlook Tower.

20. In due course Kropotkin himself became an *Encyclopaedia Britannica* contributor, writing the entry on anarchism for volume one of the eleventh edition, published in 1910.

21. Kropotkin's other Scottish friends included Keir Hardy, J. S. Keltie (assistant editor of *Nature* and later secretary of the Royal Geographical Society) and William Robertson Smith, by this time professor of Arabic at Cambridge, who floated the idea that Kropotkin should be appointed to the chair of Geography at that university (Woodcock, George and Avakumovic, Ivan, *The Anarchist Prince: A Biographical Study of Peter Kropotkin*, London: Boardman, 1950, pp. 226–9). Mavor – who shared a Free Church background with Robertson Smith (and with Geddes) – sheds light on this relationship: 'It was a matter of deep regret to me that I did not see Robertson Smith in later years . . . I used to hear of him, however, from mutual friends. He was on very intimate terms with Prince Kropotkin, and different in many ways as were the two men, there sprang up a deep mutual regard. Robertson Smith was anxious to secure Kropotkin for Cambridge as Professor of Geography. Kropotkin told me that he did not care to compromise his freedom by accepting such a position; but that he felt very pleased that Robertson Smith's friendship had prompted him to so generous a project' (Mavor, James, *My Windows*

on the Street of the World, Vol. 1., London and Toronto: Dent, 1923, p. 75).

22. Boardman, *Geddes*, pp. 210–11.

23. For the wider anarchist context of Geddes, see Ward, Colin, *Anarchism: A Very Short Introduction* (Oxford: Oxford University Press, 2004), in particular chapter 9, 'The Federalist Agenda'.

24. Swinney, Geoffrey, 'The training of a polar scientist: Patrick Geddes and the student career of William Spiers Bruce', *Archives of Natural History*, Vol. 29, No. 3, 2002, pp. 287–301, p. 291.

25. Published in 1887.

26. Mavor, 1923, *Windows*, Vol. 2, pp. 91–2.

27. Geddes, Patrick, *Co-operation versus Socialism* (Manchester: Co-operative Printing Society, 1888), p. 20.

28. For appreciation of Blackie's efforts by Gaelic poets, see Meek, Donald, *Caran an t-Saoghail: The Wiles of the World* (Edinburgh: Birlinn, 2003), pp. 218–21.

29. Grigor, Ian Fraser, *Mightier than a Lord* (Stornoway: Acair, 1979), p. 75.

30. Reilly, John P., *The Early Social Thought of Patrick Geddes*, PhD thesis (New York: Columbia University, 1972), p. 200.

31. For a wider view, see Campbell, Gordon, 'Patrick Geddes: A Regional Nationalist?', *Radical Scotland*, No. 43, 1990, pp. 28–9.

32. Patrick Geddes to J. A. C. Campbell of Barbreck, 1 November 1895. NLS MS 10508. Quoted in Cuthbert, Michael, *The Concept of the Outlook Tower in the Work of Patrick Geddes*, unpublished MPhil. thesis, University of St Andrews, 1987, pp. 194–5.

33. Ibid.

34. The Armour–Macdougall creative relationship has been explored in a study of fin-de-siècle illustrated books by Kooistra, Lorraine Janzen, *The Artist as Critic: Bitextuality in Fin-de-Siecle Illustrated Books* (Aldershot: Scolar Press, 1995), pp. 107–26.

35. *Studio*, Vol. 6, 1896, pp. 164–70, p. 165.

36. A copy of *The New Evergreen* is held by the Edinburgh Room at Edinburgh Central Library at shelfmark YLF 1047.5 (G835665).

37. The full wording of the title page is *The New Evergreen/The Christmas Book of University Hall/Published/At University Hall/Edinburgh: Christmas MDCCCXCIV*.

38. Ibid., pp. 5–6.

39. *The Evergreen: Book of Autumn* (Edinburgh: Patrick Geddes and Colleagues, 1895), p. 8.

40. Weirter, Louis, *The Story of Edinburgh Castle* (London: Harrap, 1913), p. xi. Geddes wrote his preface in Ghent in August 1913, where he was exhibiting at the Cities and Town Planning Exhibition (Exposition Comparée des Villes).

41. In 1894 Macdougall had also contributed to *The Yellow Book* and had been in direct contact with Aubrey Beardsley: Kooistra, *Artist as Critic*, p. 119, n. 33.

42. *The Artist*, April 1895, pp. 135–6.

43. Duncan does not claim this work, presumably because he regarded it as Geddes's in all but final execution.

44. Geddes, *A First Visit to the Outlook Tower* (Edinburgh: Patrick Geddes and Colleagues, 1906), p. 23.

45. Ibid., p. 25.

46. Ibid.

47. For example, McIntosh, Alastair, *Soil and Soul: People Versus Corporate Power* (London: Aurum, 2004).

48. Kooistra, *Artist as Critic*, pp. 94–107. Kooistra has a discussion of this Wilde–Ricketts collaboration that one can draw on to give insight into Duncan's iconography.

49. Reed, John, *Decadent Style* (Athens, OH: Ohio University Press, 1985), p. 14. 'Decadence is a dissolving rather than a cohering art. It is self-consciously transitional.' See also Kooistra, *Artist as Critic*, referring to Reed, p. 118, n. 2.

50. *The Evergreen Almanac*, 1896.

51. The Sphinx and the Interpreter are found in juxtaposition on the covers of Geddes's various *Masque* publications from 1912 onwards.

CHAPTER 6   Manifestos in Word and Image

1. Kemplay, John, *John Duncan – A Scottish Symbolist* (San Francisco: Pomegranate, 1994), p. 15.

2. Apted, Michael R. and Robertson, W. N., 'Late Fifteenth Century Church Paintings from Guthrie and Foulis Easter', *Proceedings of the Society of Antiquaries of Scotland*, Vol. 95, 1962, pp. 262–79. Also Dalgetty, Arthur B., *History of the Church of Fowlis Easter* (Dundee: Harley and Cox, 1933).

3. 'The Church of Fowlis Easter', *Dundee Advertiser*, 6 September 1888, in due course reprinted as a pamphlet. My thanks to Lesley Lindsay for locating this information.

4. Patrick Allan Fraser's Hospitalfield House at Arbroath (later endowed as a postgraduate art school) and its nearby mausoleum provide an earlier local example of the type of thinking that would be in due course defined as 'Arts and Crafts'. Built in the 1850s around a much older core, it provides a link between antiquarianism and the revitalisation of traditional medieval craft skills, in particular architectural wood and stone carving.

5. 'The Decorative Paintings of John Duncan', *The Artist*, 1898, pp. 146–52, p. 147.

6. Morrison, John, 'John Duncan, Patrick Geddes and the Celtic Revival', *Journal of Scottish Thought*, Vol. 5, 'Patrick Geddes', 2012, pp. 27–39.

7. A letter from Duncan to Craigie of 1885 indicates that by this date they had been friends for some years. NLS MS 9987 f.11.

8. An indication of Craigie's continuing interest in Gaelic is that he was a subscriber to Edward Dwelly's Gaelic–English dictionary, which was first issued in parts between 1902 and 1912.

9. Craigie, William A., *A Primer of Burns* (London: Methuen, 1896). For an indication of Craigie's wider achievement, see *A Memoir and a List of the Published Writings of Sir William A. Craigie* (Oxford: Clarendon Press, 1952). From the perspective of Craigie's interest in myth and legend, one can note that a subscriber to this publication was Craigie's assistant at Oxford, J. R. R. Tolkien, who followed him as Rawlinson and Bosworth Professor of Anglo-Saxon.

10. NLS MS 9987, f. 32, letter of 25 April 1893.

11. Duncan later transformed his composition as part of the background of his panel of Sir Walter Scott at Ramsay Lodge.

12. Aytoun, William Edmondstoune, *Lays of the Scottish Cavaliers* (Edinburgh: Blackwood, 1870), p. 85.

13. SUA, T-GED 9/137. Letter of 10 July 1895.

14. Armour, Margaret, 'Mural Decoration in Scotland', *The Studio*, Vol. 10, 1897, pp. 99–106.

15. Anon., 'The Decorative Paintings of John Duncan', *The Artist*, 1898, pp. 146–52.

16. Ibid., p. 147.

17. Ibid., p. 146.

18. Pater, Walter, *The Renaissance* (London: Macmillan, 1873), p. 135.

19. Cumming, Elizabeth, *Phoebe Anna Traquair* (Edinburgh: Scottish National Portrait Gallery, 1993), p. 11.

20. Ibid., p. 65.

21. SUA T-GED 9/138.

22. SUA T-GED 5/3/15. This is undated but the evidence of exhibited work points to 1896.

23. This is presumably James Oliphant, who was on the general committee of the Edinburgh Social Union.

24. Whitson became in due course Lord Provost of Edinburgh; his accountancy firm acted for University Hall.

25. *Interpretation of the Pictures in the Common Room of Ramsay Lodge* (Edinburgh: University Hall, 1944).

26. Ferguson, Megan, ' "Dear Guru" – John Duncan and Patrick Geddes', in *The Artist and the Thinker: John Duncan and Patrick Geddes in Dundee*, edited by Matthew Jarron (Dundee: University of Dundee Museum Services, 2004), pp. 3–12, p. 9. Ferguson quotes a letter from Geddes

to Duncan in which the subjects of the murals are discussed, dated 2
February 1926, NLS MS 10517.

27. Armour, Margaret, 'Mural Decoration in Scotland', *The Studio*, Vol. 10,
1897, pp. 100–6, p. 104.

28. From p. 13 of a proof copy of *The Interpreter* dated April 1896. SUA
T-GED 5/3/33.

29. Ibid., p. 14.

30. Soon after, William Hole was to make *Columba's Mission to the Picts*
the subject of a mural in the Scottish National Portrait Gallery. It is dated
1898. His fellow Edinburgh Social Union member, William McTaggart,
also painted notable Columba compositions during this period.

31. What legend this refers to is a matter of conjecture. However, the 'awakening' (i.e. reviving) is the point.

32. Given later in *The Interpreter* and subsequent guides as Fionn, but with
the explanation that 'Fionn is Macpherson's Fingal'.

33. NLS MS 10530, f. 122, 5 May 1898. To Patrick Geddes from John
Duncan, writing from 31 Albert Square, Dundee.

34. For more on this Dundee milieu, see Davidson, Senga, 'Symbolists in our
midst – Stewart Carmichael and George Dutch Davidson', in *The Artist
and the Thinker: John Duncan and Patrick Geddes in Dundee*, edited by
Matthew Jarron (Dundee: University of Dundee Museum Services, 2004),
pp. 78–90.

35. Dundee City Art Gallery.

36. Where *The Evergreen* was a design statement intended to stand the test
of time, *The Interpreter* was a low-cost, pocket-sized set of pamphlets
intended for rapid production and communication. Thomson's description of the murals can be found in the proof copy of *The Interpreter* dated
April 1896. SUA T-GED 5/3/33.

37. City Art Centre, Edinburgh.

38. NLS MS 10508A, f. 135, 18 November 1895. 'What do you say to a
picture suggested by Andrew Lang of the Franco-Scottish Society, of Joan
of Arc with her body guard of Scottish Archers?'

39. This was illustrated throughout by Selwyn Image, and printed by T. and
A. Constable in 1896. The publisher was Longmans.

40. Macdonald, Murdo, 'Anima Celtica: Embodying the Soul of the Nation in
1890s Edinburgh', in *Art, Nation and Gender: Ethnic Landscapes, Myths
and Mother-Figures*, edited by Tricia Cusack and Sìghle Bhreathnach-
Lynch (Aldershot: Ashgate, 2003), pp. 29–37. Also: Macdonald, Murdo,
'The Visual Dimension of *Carmina Gadelica*', in *The Life and Legacy of
Alexander Carmichael*, edited by Domhnall Uilleam Stiùbhart (Port of
Ness: Islands Book Trust, 2008), pp. 135–45.

41. Geddes, Patrick and Thomson, J. Arthur, 'The Moral Evolution of Sex',
in *The Evergreen: A Northern Seasonal: Book of Summer* (Edinburgh:
Patrick Geddes and Colleagues, 1896), p. 77.

42. Cf. Elisa Grilli: 'its desire to support the unity of art, science, philosophy, and *belles lettres* enhanced the dialogue between cosmopolitanism and Celtic revivalism'. Grilli, Elisa, 'Funding, Publishing, and the Making of Culture: The Case of the *Evergreen* (1895–97)', *Journal of European Periodical Studies*, 1.2, Winter 2016, pp. 19–44, p. 22.

43. Burn Murdoch, W. G., *From Edinburgh to the Antarctic* (London: Longmans, 1894), p. 31. 'Fifty years previously to this Dundee expedition of 1892–3, Sir James Ross had reported having seen great numbers of right whales in the Antarctic regions and the Dundonians having whaling vessels almost idle owing to the scarcity of right whales in the Arctic, fitted out this expedition to get these whales in the South; and our scientific bodies selected promising scientists to go as doctors on board three of the vessels and supplied them with scientific equipment.'

44. Ibid., p. 57.

45. Speak, Peter, 'Introduction' to Bruce, William Spiers, *The Log of the Scotia Expedition 1902–1904* (Edinburgh: Edinburgh University Press, 1992), p. 27. Although wanting to return to the Antarctic immediately, Bruce was unable to fund an expedition and spent a winter gaining invaluable cold-weather experience at the Ben Nevis summit observatory. In subsequent years he undertook several Arctic expeditions and in 1900, with the support of the Royal Scottish Geographical Society and Sir John Murray, he was at last able to announce what was to become the Scottish National Antarctic Expedition of 1902 to 1904. In passing, one can note that another Scottish visual thinker who used the Ben Nevis summit observatory during this period was C. T. R. Wilson. In his Nobel lecture from 1927, Wilson recalls his stay at the observatory in September 1894 as central to research underlying his invention of the Cloud Chamber (the key visual tool of particle physics from 1912 until at least the mid-1950s). He also visited the summit at the time of Bruce's association with the observatory in summer 1895; see Swinney, Geoffrey, 'William Spiers Bruce, the Ben Nevis Observatory, and Antarctic Meteorology', *Scottish Geographical Journal*, Vol. 118 (4), 2002, pp. 263–82, p. 270. Another of Bruce's visitors at the summit was the geologist Archibald Geikie.

46. Rudmose Brown, R. N., assisted by W. G. Burn Murdoch, *A Naturalist at the Poles: The Life, Work and Voyages of Dr W. S. Bruce* (London: Seeley Service and Co., 1923), p. 296. The Scottish National Antarctic Expedition can be followed from inception to achievements through the pages of the *Scottish Geographical Magazine* from 1900 onwards.

47. Ibid., p. 31.

48. Rudmose Brown, *A Naturalist at the Poles*, pp. 32–3.

49. Ibid., p. 27.

50. Ibid., pp. 295–6.

51. Willsdon, Clare, 'Paul Sérusier the Celt: did he paint murals in Edinburgh?', *The Burlington Magazine*, Vol. 126, 1984, pp. 88–91.

52. Cf. J. G. Fletcher's emphasis on Brittany as part of the Celtic fringe of Europe in *Paul Gauguin: His Life and Art* (New York: Brown, 1921), p. 55.

53. Willsdon, Clare A. P., *Mural Painting in Britain 1840–1940: Image and Meaning* (Oxford: Oxford University Press, 2000), pp. 326–7. For wider context, see Willsdon, Clare, 'The Ramsay Garden Murals and their Links with French Mural Painting', *Journal of the Scottish Society for Art History*, Vol. 9, 2004, pp. 69–78.

54. For a considered assessment of the validity of this attribution, see G. Ross Roy, '*Hardyknute* – Lady Wardlaw's Ballad?', in *Romanticism and Culture*, edited by H. W. Matalene (Columbia: Camden House, 1984), pp. 134–46.

55. In 1922 Macgillivray made a selection of his poems under the title *Bog Myrtle and Peat Reek*. Here his interest in both Highland and Lowland is explicit. The book itself is a cultural statement, for it takes the model for its production from the materials, design and typeface of the Kilmarnock edition of Burns's poems.

56. It is interesting to note that among Geddie's other publications was the text for *Illustrations of the Scenery of the River Tay*, published in 1891. This was printed, like *The Evergreen*, by T. and A. Constable and illustrated in part by *Evergreen* artist Charles Mackie. The text is notable for including a comparison of the Tay and the Ganges. Geddie's *The Fringes of Fife*, 1894, was illustrated by *New Evergreen* contributor Louis Weierter.

57. Miller was another notable Scottish generalist figure of the mid-nineteenth century. A stonemason by trade, he became a significant geologist. At the same time he was a talented journalist who wrote one of the earliest accounts of the calotype process of photography. He was also a central figure in the formation of the Free Church of Scotland in 1843.

58. It was in this series that Elizabeth Haldane was later to publisher her book on the philosopher James Frederick Ferrier. One can also note that the volume devoted to Robert Burns was written by *Evergreen* contributor Gabriel Setoun.

59. Geddie, John, *The Balladists* (Edinburgh: Oliphant, Anderson and Ferrier, 1896), p. 144. For an informed assessment of the significance of this earlier work for Geddes and his colleagues, see Elliot, Elizabeth, 'Old-World Verse and Scottish Renascence: Flourishing Evergreen', in *The Evergreen: A New Season in the North* (Edinburgh: The Word Bank, 2014), pp. 149–56.

60. *The Evergreen: Book of Autumn*, p. 8.

61. *The Evergreen: Book of Spring*, pp. 16–17.

62. Still referred to as *The New Evergreen*, reviewed in *The Artist*, 1895, pp. 270–3.

63. *The Evergreen*, reviewed in *The Artist*, 1896, pp. 519–20.

64. Walter Crane, *Of the Decorative Illustration of Books* (London: George Bell, 1896), p. 227.

65. Ibid., pp. 255–7.

66. SUA T-GED 8/1/8. The letter is contained in a volume of newspaper cuttings which contains about one hundred reviews of *The Evergreen*.

67. Crane, *Decorative Illustration of Books*, p. 187.

68. Ibid., p. 186.

69. SUA T-GED 9/138.

70. *An Rosarnach* (Glascu/Glasgow: Alasdair Mac Labhrainn, 1930).

71. *The Studio*, Vol. 13, 1898, pp. 48–9.

72. See MacCormick, Iain, *The Celtic Art of Iona* (Iona: New Iona Press, 1994), MacArthur, E. Mairi, *Iona Celtic Art: The Work of Alexander and Euphemia Ritchie* (Iona: New Iona Press, 2003) and Elizabeth Cumming, *Hand, Heart and Soul: The Arts and Crafts Movement in Scotland* (Edinburgh: Birlinn, 2006), pp. 178–80.

73. The society was founded by the eleventh Earl of Buchan in 1780. Buchan's polymathic approach has something in common with Geddes a century later. In the mid-nineteenth century the society owed a great deal to the antiquarian activities of artists of an older generation than John Duncan, such as Joseph Noel Paton, James Drummond and William Fettes Douglas.

74. Anderson, Joseph, 'Notice of a bronze sword, with handle-plates of horn, found at Aird, in the Island of Lewis', *Proceedings of the Society of Antiquaries of Scotland*, Vol. XXVII, 1893, pp. 38–41. My thanks to Finlay Macleod of Shawbost for drawing my attention to this paper by Anderson and its possible relation to Duncan's work.

75. In the *Celts* exhibition held at the National Museum of Scotland in 2016 an alternative source was suggested, namely a sword found at Leadburn. This suggestion is of value, for Duncan's version seems to incorporate features from both finds, the horn handle of the Aird find and the rounded pommel of the Leadburn find.

76. Cheape, Hugh, 'The Culture and Material Culture of Ossian', *Scotlands*, 4.1, 1997, pp. 1–24, p. 15.

77. I use the various spellings given at the time.

78. These objects can now be found in the Museum of Scotland in Chambers Street.

79. Comrie, Duncan, 'Cultural Consciousness in Edinburgh', *History Scotland*, Vol. 2, No. 4, 2002, pp. 12–21. Comrie comments on p. 20 that the 'art establishment . . . excluded from the SNPG murals any explicit Celtic cultural identity'. This may overstate the case, but the point is an interesting one.

80. For a sense of the scale of these celebrations, see 'The Centenary of the Death of Burns', in *The National Burns* (Glasgow: William Mackenzie, n.d., c. 1896), pp. 345–86.

81. For example, Geddes, Patrick, *The Masque of Medieval and Modern Learning* (Edinburgh: Patrick Geddes and Colleagues, 1913), p. 47.
82. MacDiarmid, Hugh, *Robert Burns* (London: Ernest Benn, n.d., c. 1925), p. iii.

CHAPTER 7   Models of Celticism

1. The only exception to the ban was for soldiers serving in Scottish regiments of the British army.
2. Graham, Henry Grey, *Scottish Men of Letters in the Eighteenth Century* (London: Adam and Charles Black, 1901).
3. Thomson, Derick, *The Gaelic Sources of Macpherson's Ossian* (Edinburgh: Oliver and Boyd, 1951). Thomson's work at the University of Glasgow began the process of rethinking Macpherson taken further from the 1980s onwards by *inter alia* Fiona Stafford and Howard Gaskill. See, in particular, Stafford, Fiona, *The Sublime Savage: A Study of James Macpherson and the Poems of Ossian* (Edinburgh: Edinburgh University Press, 1988); and Gaskill, Howard, ed., *The Reception of Ossian in Europe* (London: Continuum, 2004).
4. Sharp, Elizabeth A., *William Sharp (Fiona Macleod): A Memoir* (London: Heinemann, 1910), pp. 249–55.
5. Rhys was also an important editor. From 1886 he edited the Camelot series for Walter Scott, and in 1905 went on to name and edit the Everyman series for J. M. Dent. See Rhys, Ernest, *Everyman Remembers* (New York: Cosmopolitan Book Corporation, 1931).
6. Alaya, Flavia, *William Sharp – 'Fiona Macleod'* (Cambridge, MA: Harvard University Press, 1970), pp. 124–6.
7. Graves, Alfred Perceval, *The Book of Irish Poetry* (London: Unwin, 1915), p. xxxviii.
8. Given as 'The Lake of Innisfree'.
9. Sharp, William, 'Introduction', *Lyra Celtica*, edited by Elizabeth A. Sharp (Edinburgh: Patrick Geddes and Colleagues, 1896), p. li. Sharp's subjugation of Gaelic to English mirrored the continuing destruction of Gaelic-speaking communities in Scotland, in the wake of the anti-Gaelic Education Act of 1872.
10. Ibid., p. xliv.
11. Yeats, William Butler, *The Collected Letters of W. B. Yeats Volume 2*, edited by Warwick Gould, John Kelly and Deirdre Toomey (Oxford: Oxford University Press, 1996), p. 56, footnote.
12. Carmichael, Alexander, 'The Land of Lorne and the Satirists of Taynuilt', *The Evergreen: Book of Spring* (Edinburgh: Patrick Geddes and Colleagues), pp. 110–15.
13. Reprinted as 'The Gael and his Heritage' in *The Winged Destiny, Studies*

in the Spiritual History of the Gael, Vol. 5 of the Uniform Edition of the works of Fiona Macleod.

14. Carmichael, Alexander, *Carmina Gadelica*, Vol. 1 (Edinburgh: Printed for the Author by T. and A. Constable and sold by Norman MacLeod), p. xxxii.
15. Alaya, *Sharp*, pp. 148–51.
16. Ibid., p. 148.
17. MacInnes, John, 'New Introduction to the *Carmina Gadelica*' (1992), in *Dùthchas Nan Gàidheal: Selected Essays of John MacInnes*, edited by Michael Newton (Edinburgh: Birlinn, 2006), pp. 467–92.
18. Ibid., p. 492. MacInnes is quoting from Raghnall MacIlleDhuibh (Ronald Black), 'The Trouble with *Carmina*', *West Highland Free Press*, 29 May 1992.
19. Given the interdisciplinary standpoint of this book, it can be noted that Sorley MacLean was introduced to Hugh MacDiarmid by George Davie. See Macdonald, Murdo, 'George Davie: Life and Significance', in Davie, George, *The Democratic Intellect* (Edinburgh: Edinburgh University Press, 2013), pp. x–xiii, p. xi.
20. Carmichael, *Carmina*, Vol. 1, p. xxxii.
21. Ibid., p. xxxi–ii.
22. My thanks to Iain Maciver of the National Library of Scotland for guiding me to the correct body of manuscripts and drawing my attention to Ronald Black's identification of initials used as models by Mary Macbean.
23. See Macdonald, Murdo, 'The Visual Dimension of *Carmina Gadelica*', in *The Life and Legacy of Alexander Carmichael*, edited by Domhnall Uilleam Stiùbhart (Port of Ness: Islands Book Trust, 2008), pp. 135–45; also Macdonald, Murdo, 'The Visual Preconditions of Celtic Revival Art in Scotland', *Journal of the Scottish Society for Art History*, Vol. 13, *Highlands* issue, 2009, pp. 16–21.
24. Carmichael uses the spelling 'Deirdire'.
25. *Deirdire and The Lay of the Children of Uisne Orally Collected in the Island of Barra, and Literally Translated by Alexander Carmichael* (Edinburgh: Norman MacLeod, 1905). The acknowledgement of Duncan's gift does not appear until the second edition (Paisley: Alexander Gardner, 1914).
26. SUA T-GED 12/3/31. John Duncan to Patrick Geddes from 29 St Bernard's Crescent, Edinburgh. Probably Spring 1912.
27. Macdonald, *Visual Dimension*, p. 142.
28. Dwelly also included an appendix of names from Macpherson's *Ossian*, derived from *Armstrong's Gaelic Dictionary* (London: James Duncan, 1825).
29. Marjory Kennedy-Fraser, 1929, *A Life in Song* (London: Oxford University Press, 1929), p. 198.
30. Ibid., p. 4.
31. MacLean, Sorley, 'Realism in Gaelic Poetry', collected in Somhairle

Mac Gill-eain, *Ris a Bhruthaich: Criticism and Prose Writings of Sorley MacLean*, edited by William Gillies (Stornoway: Acair, 1985), pp. 15–47, p. 20.

32. John Purser has pointed out that her work should be seen in the context of the contemporary adaptation of folk material by Debussy and Bartok. Purser, John, *Scotland's Music* (Edinburgh: Mainstream, 1992), p. 241.
33. *Our Singer and Her Songs* (Edinburgh: Patrick Geddes and Colleagues, 1931).
34. MacLeod, Kenneth, *The Road to the Isles: Poetry, Lore and Tradition of the Hebrides* (Edinburgh: Grant, 1927).
35. Kennedy-Fraser, *Life*, pp. 78–9.
36. Ibid., p. 120.
37. Edinburgh: Patrick Geddes and Colleagues; Dublin: Gill and Son, and E. Ponsonby.

CHAPTER 8   Interdisciplinarity and Cultural Revival

1. 'The Edinburgh Summer Meeting', *The Evergreen Almanac* (Edinburgh: Patrick Geddes and Colleagues, 1896), final page (unnumbered). SUA T-GED 8/1/7.
2. Edinburgh Summer Meeting University Hall August 3–29, 1896 (Part I. Aug. 3–15) (Part II. Aug. 17–29). SUA T-GED 7/8/23.
3. The cover also states: 'To be had from T. R. Marr, Outlook Tower, University Hall, Edinburgh Price Fourpence March 1896'.
4. Sharp's contribution to the event was curtailed by illness.
5. Sharp, Elizabeth, *William Sharp: A Memoir* (London: Heinemann, 1910), p. 250.
6. Kennedy-Fraser, Marjory, *A Life of Song* (London: Oxford University Press, 1929), p. 120.
7. *Prospectus* inside back cover. *The Interpreter*, Nos. 1–24, Tenth Edinburgh Summer Meeting, August 1896. Copy held in Edinburgh Central Library, Edinburgh Room, YLF 1047.5U (18904).
8. *Interpreter*, No. 8, 10 August 1896, p. 3. The writer is anonymous, but I have presumed it to be Geddes.
9. Proof copy of *The Interpreter* dated April 1896. SUA T-GED 5/3/33.
10. See drawing in the Strathclyde University archive, at SUA T-GED 5/2/13. I note again here that the developed notation or chart of life was published in Defries, Amelia, *The Interpreter: Geddes, the Man and his Gospel* (London: Routledge, 1927), pp. 123–44. Its pattern, but with differently positioned terminology, is discussed in Geddes, Patrick, *The World Within and the World Without: Sunday Talks with my Children* (Bournville: Saint George Press, 1905). For its use in an environmental architectural context, see Sato, Fumiaki, *A Critical Study of Housing and*

Sustainability: A Japanese Exemplar, unpublished PhD thesis, University of Edinburgh, 2003.

11. A work by Margaret Macdonald from 1894 is suggestive of more direct links with Geddes's milieu at this time. It is a book cover design for *The Story of a River*. See Howarth, Thomas, *Charles Rennie Mackintosh and the Modern Movement* (London: Routledge & Kegan Paul, 1952), plate 6D. The title may have found inspiration in Reclus's essay *Historie d'un Ruisseau* (Paris, 1869), an essay that underpins Geddes's concept of the valley section.

12. Comenius, John Amos, *Orbis Pictus, A facsimile of the first English edition of 1659* (London: Oxford University Press, 1968).

13. Laurie, Simon Somerville, *John Amos Comenius* (London: Kegan Paul, Trench and Co., 1881), p. 191.

14. Those present at the opening ceremonies of the new premises of the Scottish Arts Club in Rutland Square, Edinburgh (which coincided with the election of lay members) included, as well as Laurie and Masson, a number of those involved in the Edinburgh Social Union, for example Gerard Baldwin Brown, George Reid, D. W. Stevenson and W. B. Blaikie. The opening was conducted by the Manx cultural revivalist Hall Caine. (The Scottish Arts Club: An Account of the Opening Ceremonies and of the speeches delivered at the opening dinner on Friday Ninth November 1894, Edinburgh, Privately Printed). Geddes is not mentioned here but is noted as having become a member in 1894. See *The Scottish Arts Club 1874–1974* (Edinburgh: The Scottish Arts Club, 1974), p. 48.

15. *University Hall, Edinburgh* (Edinburgh: Privately Published, 1903). The quotation is taken from the back cover of this pamphlet, a copy of which is held by the Edinburgh Room of the Edinburgh Central Library at shelfmark YLF 1047 5.

16. Armour, Margaret, 'Mural Decoration in Scotland', *The Studio*, Vol. 10, 1897, pp. 100–6, p. 106.

17. 'The Edinburgh Summer Meeting', *Evergreen Almanac*, final page (unnumbered). The piece is unsigned. Geddes is the probable sole author, but joint authorship with Thomson or Branford or both is possible.

18. Ibid.

19. Ibid.

20. Boardman, Philip, *The Worlds of Patrick Geddes* (London: Routledge, 1978), p. 129.

21. Zueblin, Charles, 'The world's first sociological laboratory', *American Journal of Sociology*, Vol. 4, No. 5, 1899, pp. 577–91. Heinz Maus comments in his *A Short History of Sociology* (London: Routledge & Kegan Paul, 1962), that 'one may say that it is with Geddes that the town first moves into the purview of sociology, which, astonishing as it may seem, had up to then dealt with it only casually and occasionally' (p. 47). In

1922 Lewis Mumford called the Outlook Tower the point of origin of the Regional Survey Movement: Mumford, Lewis, *The Story of Utopias* (New York: Boni and Liveright, 1922).

22. Personal communication. My thanks to Annie V. F. Storr, Women's Studies Research Center, Brandeis University for allowing me to report her research. My debt is also to Daniel Meyer, Archivist of the University of Chicago. The key letter is from Geddes to Harper, dated 16 November 1891, and can be found in the archives of the University of Chicago in the William Rainey Harper Papers, Box XIII, Folder 32.

23. *The Civic Survey of Edinburgh* (Edinburgh: Outlook Tower, 1911), reprinted from *The Transactions of the Town Planning Conference*, 1911. Reprinted again with modifications as 'Beginnings of a Survey of Edinburgh', in *The Scottish Geographical Magazine*, Vol. 25, 1919, pp. 281–98.

24. 'The Edinburgh Summer Meeting', *Evergreen Almanac*, final page (unnumbered).

25. Ibid. See also John Kelman's 'The Education of the Eye', an essay based around Geddes's activities at the Outlook Tower. The essay was a contribution to a volume sold in aid of the Royal Victoria Hospital on the occasion of the Edinburgh International Fair of November 1905. The book, *A Beggar's Wallet*, was edited by John Stuart Blackie's son-in-law, A. Stodart Walker. It also contained images by John Duncan, James Cadenhead and other artists of Geddes's circle. Writers included Gabriel Setoun, John Buchan and Rosaline Masson. Kelman's essay *The Interpreter's House: The Ideals embodied in the Outlook Tower, Edinburgh*, had been published the same year.

26. 'The Outlook Tower: General Plan and Arrangement', *Evergreen Almanac*, final page (unnumbered).

27. Ibid.

28. Ibid.

29. 'Hence the real comparison of Edinburgh and Athens – each plainly a hill fort at once with a seaport and with an agricultural plain . . . Thus we see the traditional comparison of Edinburgh and Athens has really little to do with our eighteenth and nineteenth century imitations of Greek temples or Greek sophistries, but lies far deeper, in geographical and historical origins': Geddes, *Civic Survey*, p. 283.

30. Geddes, Patrick, 'The Index Museum: Chapters from an Unpublished Manuscript', *Assemblage*, No. 10, December 1989, pp. 65–9. Appended to Ponte, Alessandra, 'Building the Stair Spiral of Evolution: The Index Museum of Patrick Geddes', *Assemblage*, No. 10, December 1989, pp. 47–69.

31. NLS MS 10511 f100. Patrick Geddes to Dr Paton, 7 February 1905, writing from 6 Christchurch Road, Hampstead. See also Boyer, M. C., *The City of Collective Memory* (Cambridge, MA: MIT Press, 1994), pp. 221–3, for

comment on Geddes's inspiration in Diderot and d'Alembert's approach to visual material.

32. *A First Visit to the Outlook Tower* (Edinburgh: Geddes and Colleagues, 1906), p. 21, p. 23, p. 26.
33. SUA T-GED 8/1/3.
34. The actual wording is 'Stammbaum as Aloe Plicatilis' and then in brackets the words 'Evergreen cover' and 'cf. Arbor Saeculorum'.
35. NLS MS 10508a f 94.
36. *First Visit*, p. 26.
37. Ibid., p. 28.
38. Ibid., pp. 28–9.
39. Ibid., p. 29.
40. Ibid. pp. 21–2.
41. SUA T-GED 7/2/6.
42. Geddes uses the Scottish term 'crofter' in the version of the valley section reproduced on p. 73 of Amelia Defries's *The Interpreter: Geddes*. [Figure 8.]
43. NLS MS 10509, f. 111–15a. Quote is from f. 112.
44. In this letter Geddes translates *famille* as 'family' rather than using his later more well-known adaptation of it to 'folk'.
45. In other versions of the diagram, for example one in the archives of the University of Dundee (University of Dundee Archive Services MS 197/9 Geddes's valley plan, n.d.) [Figure 7], Geddes included the term 'kakatopia' as an alternative to 'eutopia', showing that he was keenly aware that human decision-making, while having the potential to create a 'good place' of the future, had, equally, the potential to create the opposite.
46. Maker unknown. The doves also appear on the ceiling of the dining room of Riddle's Court painted to Thomas Bonnar's design in 1897. They also appear on a stained-glass window in another University Hall residence, St Giles House.
47. Burns, Robert, 'Epistle To James Tennant Of Glenconner', 1789.
48. The immediate source for Geddes of this triad of ideas may have been Henry Maudsley's essay *Body and Will*, which had been published in 1883. Maudsley, Henry, *Body and Will* (London: Kegan, Paul, Trench and Co.,1883), pp. 302–3. My awareness of this text is due to Roger Luckhurst's inclusion of the relevant excerpt as an appendix to his edition of Robert Louis Stevenson's *Strange Case of Dr Jekyll and Mr Hyde* (Oxford: Oxford University Press, 2006). The analogies between Geddes and Stevenson deserve further exploration, both in terms of their shared interest in biological evolution and in terms of their mutual advocacy of cultural pluralism.
49. *Evergreen: Book of Summer* (Edinburgh: Patrick Geddes and Colleagues, 1896), pp. 43–63.
50. First Epistle of Peter, verses 24–5, *Authorised Version of the Bible*.

51. *First Visit*, p. 23
52. Geddes, *Flower of Grass*, p. 44.
53. Ibid., p. 58. Geddes gives a more positive appreciation of Comte in his introduction to Susan Liveing's biography of her uncle, John Henry Bridges, published in 1926.
54. Paul Reclus's stay in Edinburgh was as a result of threatened imprisonment in France due to his anarchist politics. He stayed in the Outlook Tower under the assumed name of 'Georges Guyou', under which name he published an account of the Dreyfus case, introduced and published by Geddes.
55. Arthur Geddes gives a sense of this approach to the scale and curvature of the planet in the introductory image to his paper 'The "Outer" Hebrides', *The New Naturalist*, Summer 1948, pp. 72–6.
56. Boardman, *Geddes*, p. 194.
57. Defries, *Geddes*, p. 188.
58. Davie, Elspeth, *Coming to Light* (London: Hamish Hamilton, 1989), p. 150.
59. It is appropriate that these proposals can now be found in the National Library of Scotland itself. NLS MS 10616, ff 52–65. 'Extract from Colleges and Their Making' – 'National Library of Scotland'. Pencilled date of 1912 on folio 52.

CHAPTER 9   Paris 1900

1. Boardman, Philip, *The Worlds of Patrick Geddes* (London: Routledge, 1978), p. 153. Cf. Lewis Mumford's comment: 'Geddes's Scotland embraced Europe, and his Europe embraced the world'. *The Condition of Man* (London: Secker and Warburg, 1944), p. 383.
2. Davie, George, *The Democratic Intellect* (Edinburgh: Edinburgh University Press, 1961), p. 333.
3. Robertson, Stewart, *A Moray Loon* (Edinburgh: Moray Press, 1933), p. 2.
4. Willsdon, Clare A. P., 'Paul Sérusier the Celt: did he paint murals in Edinburgh?', *The Burlington Magazine*, Vol. 126, 1984, pp. 88–91.
5. Translated as *The Night-Comers*. Sharp also includes an extensive note on Lerberghe, noting him as the predecessor of Maeterlinck.
6. Cf. J. G. Fletcher's emphasis on Brittany as part of the Celtic fringe of Europe in his 1921 biography of Gauguin: *Paul Gauguin: His Life and Art*, p. 55. For discussion of Mackie's Pont-Aven links, see Thomson, Belinda, 'Patrick Geddes's "Clan d'Artistes": some elusive French connections', in *Patrick Geddes: The French Connection*, edited by Frances Fowle and Belinda Thomson (Oxford: White Cockade, 2004), pp. 47–68.
7. Willsdon, Clare A. P., *Mural Painting in Britain 1840–1940: Image and Meaning* (Oxford: Oxford University Press, 2000), pp. 326–7.
8. These lithographs, under the title *St. Geneviève of Paris*, are advertised

as part of Geddes's *Ethic Art Series*, on the back cover of *The Evergreen Anthology* (Edinburgh: Patrick Geddes and Colleagues, c. 1897). The reference to the prints forms part of a general advertisement for the publications of Patrick Geddes and Colleagues, which includes also publicity for *The Evergreen* and the *Celtic Library*, including Edith Wingate Rinder's retellings of Breton legends under the overall title *The Shadow of Arvor*.

9. Now in the collection of the City Art Centre, Edinburgh.

10. NLS MS 10508A, f. 135. Patrick Geddes to John Duncan, 18 November 1895: 'What do you say to a picture suggested by Andrew Lang of the Franco-Scottish Society, of Joan of Arc with her body guard of Scottish Archers?'.

11. *The Evergreen: Book of Summer* (Edinburgh: Patrick Geddes and Colleagues, 1896), p. 98. For context, see Reynolds, Siân, 'Patrick Geddes's French Connections in Academic and Political Life: Networking in France from 1878 to the 1900s', in *Patrick Geddes: The French Connection*, edited by Frances Fowle and Belinda Thomson (Oxford: White Cockade, 2004), pp. 69–82.

12. Lang, Andrew, *A Monk of Fife* (London: Longmans, 1896). The printer was T. and A. Constable.

13. Ponte, Alessandra, 'Building the Stair Spiral of Evolution: The Index Museum of Patrick Geddes', *Assemblage*, No. 10, December 1989, pp. 47–69, p. 56.

14. Geddes, Patrick, *Industrial Exhibitions and Modern Progress* (Edinburgh: David Douglas, 1887).

15. Mavor, James, *My Windows on the Street of the World* (London and Toronto: Dent, 1923), Vol. 2, p. 107.

16. Reynolds, 'Geddes's French Connections', pp. 78–80.

17. Steele, Tom, 'Élisée Reclus and Patrick Geddes: Geographies of the Mind', 2002. http://www.haussite.net/haus.0/SCRIPT/txt2000/04/reclus_geddes. HTML

18. Ibid.

19. James Mavor points out that its full title was 'International Association for the Advancement of Science, Arts and Education – First Assembly at the Paris Exposition of 1900': *My Windows*, Vol. 2, p. 107. In 1899 Geddes had persuaded a wealthy Perth businessman, Sir Samuel Pullar, to donate funds to make this International Association possible.

20. Addams, Jane, 'A Visit to Tolstoy', *McClure's Magazine*, Vol. 36, January 1911, pp. 295–302. This reference, complete with John Duncan's image, was drawn to my attention by Annie V. F. Storr, Women's Studies Research Center, Brandeis University.

21. Wright, Frank Lloyd, *The Art and Craft of the Machine* (1901), in *Frank Lloyd Wright, Collected Writings Volume I*, edited by B. B. Pfeiffer (New York: Rizzoli, 1995).

22. This provides background for the comment made by David Daiches,

derived from his time teaching at the University of Chicago, that Geddes was an influence on Frank Lloyd Wright. Daiches was speaking on the Radio Scotland programme *Patrick Geddes*, produced and directed by Billy Kay, September 1996. The link is illuminated by the friendship of both Geddes and Wright with C. R. Ashbee. Note also Lewis Mumford's comment that Frank Lloyd Wright 'was one of the handful of people I have known who, through the direct impact of their personalities, I would place at the same level as Patrick Geddes', quoted in Meehan, Patrick J., *Frank Lloyd Wright Remembered* (Washington: The Preservation Press, 1991), p. 200.

23. NLS MS 10511 f. 61.
24. Annie V. F. Storr made me aware of the importance of the link between John Duncan and Ellen Gates Starr.
25. Quoted from SUA T-GED 6/1/6 by Siân Reynolds in 'After Dreyfus and before the Entente: Patrick Geddes's Cultural Diplomacy at the Paris Exhibition of 1900', in *Problems in French History*, edited by M. Cornick and C. Crossley (London: Palgrave Macmillan, 2000).
26. Reynolds, 'Geddes's French Connections', p. 69.
27. Mavor, *My Windows*, Vol. 2, p. 111.
28. It is unlikely that Kropotkin himself was present, since in 1900 he was still unwelcome in France. My thanks to Siân Reynolds for pointing this out.
29. Mavor, *My Windows*, Vol. 2, pp. 119–20.
30. Iyer, M. Srinivasa, 'Sister Nivedita', in *Eminent Orientalists: Indian, European, American* (Madras: G. A. Nateson and Co., 1922), pp. 257–82, p. 259.
31. Brockington, J. L., *The Sacred Thread: A Short History of Hinduism*, second edition (New Delhi: Oxford University Press, 1997), p. 182.
32. Guha-Thakurta, Tapati, *The Making of a New 'Indian' Art: Artists, Aesthetics and Nationalism in Bengal, c. 1850–1920* (Cambridge: Cambridge University Press, 1992), p. 173.
33. Nivedita, Sister, *Kali the Mother* (London: Swan Sonnenschein, 1900).
34. Patrick Geddes, in S. K. Radcliffe, ed., *Margaret Noble (Sister Nivedita)* (London: Sherratt and Hughes, 1913). There is a copy in the Strathclyde University Archive at SUA T-GED 3/3/35. It was originally part of a two-paper feature on Nivedita in *The Sociological Review*, Vol. 13, pp. 242–56.
35. Iyer, 'Nivedita', p. 257.
36. Geddes, 'Nivedita', p. 12.
37. Ibid.
38. Ibid., p. 13.
39. Nivedita, Sister, epigraph, *The Web of Indian Life* (London: Heinemann, 1904).
40. For the links between Okakura and India, see Guha-Thakurta, *New Indian Art*, pp. 167–71, and Mitter, Partha, *Art and Nationalism in Colonial*

India (Cambridge: Cambridge University Press, 1994), pp. 262–6. Also Macdonald, Murdo, 'Patrick Geddes and Cultural Renewal through Visual Art: Scotland–India–Japan', in *Patrick Geddes: By Leaves We Live* (Yamaguchi: Yamaguchi Institute of Contemporary Art, 2005), pp. 46–72.

41. Okutsu, Kiyoshi, 'Aesthetics of the Meiji Era and Geddesian Thought', in *Patrick Geddes: By Leaves We Live* (Yamaguchi: Yamaguchi Institute of Contemporary Art, 2005), pp. 11–28.
42. Okakura, Yoshisaburo, *The Japanese Spirit* (London: Constable, 1905).
43. Okakura, Kakuzoa, *The Book of Tea* (New York: Putnam, 1906).
44. Ando, Toshihiko, 'Quest for an Alternative Outlook: Geddes and Japan, 1900–2004', in *Patrick Geddes: By Leaves We Live* (Yamaguchi: Yamaguchi Institute of Contemporary Art, 2005), pp. 72–83, p. 79.
45. Okutsu, 'Aesthetics', pp. 16–17.
46. Bradley, Jane, 'How Scotland lit up star Japanese writer's life', *The Scotsman*, 22 September 2018.
47. Flanagan, Damian, 'Century-old postcard sheds light on dark days of author Soseki's life in UK', *Mainichi* (Japan), 15 June 2018.
48. A further possible point of contact between Soseki and Geddes is the Paris Exhibition of 1900. Soseki had travelled from Japan to London via the exhibition.
49. Schwarzer, Mitchell, 'Cosmopolitan Difference in Max Dvořák's Art Historiography', *The Art Bulletin*, Vol. 74, No. 4, 1992, pp. 669–78, p. 678.
50. Ibid., pp. 669–70.
51. Neat, Timothy, *Part Seen, Part Imagined: Meaning and Symbolism in the Work of Charles Rennie Mackintosh and Margaret Macdonald* (Edinburgh: Canongate, 1994), p. 19.
52. See also Rampley, Matthew, 'Max Dvořák: art history and the crisis of modernity', *Art History*, Vol. 26, No. 2, April 2003, pp. 214–37.
53. On this point, see my 'Finding Scottish Art' in *Across the Margins: Cultural identity and change in the Atlantic archipelago*, edited by G. Norquay and G. Smyth (Manchester: Manchester University Press, 2002), pp. 171–84.

CHAPTER 10 City Development as Interdisciplinary Project

1. *A Study in City Development: Parks, Gardens and Culture-Institutes.*
2. NLS MS 10538, ff. 67–70.
3. The wording is as follows: 'A report to the Carnegie Dunfermline Trust by Patrick Geddes, Professor of Botany, University College Dundee (St Andrews University), President of the Edinburgh School of Sociology'.
4. NLS MS 10534, ff. 29, 35. Letter dated 16 February 1903.
5. For a contemporary account of these early developments in sociology, see Saleeby, C. W., *Sociology* (London: Jack, n.d.). The book is undated

but, from internal evidence, must have been written around 1912 or 1913.

6. *University of London: The Historical Record 1836–1912* (London: Hodder and Stoughton), pp. 184–7.

7. Geddes, *City Development*, pp. 140–1 and pp. 145–7.

8. Ibid., p. 199.

9. Ibid., p. 145.

10. Ibid., pp. 144–5. An example of Burn Murdoch's frieze in its paper form can be seen in the schoolroom at Innerpeffray Library in Perthshire.

11. Ibid., p. 145. The artists were John Lavery, Alexander Roche, David Forrester Wilson and George Henry.

12. Hubbard, Tom, 'Patrick Geddes: Benign Faust', *Edinburgh Review*, No. 88, 1992, pp. 51–7, p. 53.

13. Over a decade after Geddes's death, Mumford would assert this inspiration in a book of his own essays which took the same name as the Dunfermline report, namely 'City Development'. Mumford, Lewis, *City Development: Studies in Disintegration and Renewal* (New York: Harcourt, Brace, 1945). Seven years after he encountered the Dunfermline report and a year before he met Geddes face to face, Mumford's first book, *The Story of Utopias*, was published (New York: Boni and Liveright, 1922). It contains an account of Geddes's thinking particularly with reference to his ideas of regional survey.

14. Mumford, Lewis, *The Condition of Man* (New York: Harcourt, Brace, 1944), p. 384.

15. Jarron, Matthew, 'Patrick Geddes and Museum Ideas in Dundee and Beyond', *Museum Management and Curatorship*, No. 21, 2006, pp. 88–94, p. 90.

16. In this decade, the remarkable I. F. Grant started her collection of Highland and Hebridean artefacts in a disused church on the island of Iona. This led to the establishment of the open-air Highland Folk Museum at Kingussie in 1944. See Grant, I. F., *Highland Folk Ways* (London: Routledge & Kegan Paul, 1961), and Grant, I. F., *The Making of* Am Fasgadh: *An Account of the Origins of the Highland Folk Museum by its Founder* (Edinburgh: National Museums of Scotland, 2007).

17. By this time Geddes had published his extended essay 'Civics as Applied Sociology' in 1905 and 1906 in *Sociological Papers*. For accounts of the history of the Sociological Society, see Halliday, R. J., 'The Sociological Movement, The Sociological Society and the Genesis of Academic Sociology in Britain', *The Sociological Review*, New series, Vol. 16, No. 3, 1968, pp. 377–98, and Evans, D. F. T., *Le Play House and the Regional Survey Movement in British Sociology 1920–1955*, MPhil. thesis, City of Birmingham Polytechnic, 1986.

18. Lewis Mumford quoting Patrick Geddes and Victor Branford in *The Culture of Cities* (New York: Harcourt, Brace, 1938), p. 6.

19. Ibid., pp. 6–7.
20. Geddes, Patrick, *Cities in Evolution* (London: Williams and Norgate, 1915), p. 13.
21. Kraus, Sabine, 'Aristotélisme, darwinisme et holisme chez Patrick Geddes', *Espaces et sociétés* 167, n° 4/2016, pp. 121–35, p. 132.
22. Geddes's essay *The Civic Survey of Edinburgh* is a useful visual complement to *Cities in Evolution*. It was published in 1911 complete with images of speculative reconstructions of the city at different dates by his future son-in-law Frank Mears. This formed the basis for essays by both Geddes and Mears in the 'Edinburgh' number of *The Scottish Geographical Magazine* published in 1919. Both are good examples of the kind of high density of comparative visual material Geddes advocated in his exhibitions: complementing the essays was a selection of historical views and maps given coherence by J. G. Bartholomew's 'Chronological Map of Edinburgh, showing Expansion of the City from Earliest Times to the Present'.
23. Geddes, Patrick, *Cities in Evolution* (London: Williams and Norgate, 1915), pp. 368–9. He continues with the words I have quoted in chapter one: 'Among the educators of history there are few more significant and perhaps none at this moment more vividly modern, more directly indicative of the twofold needs of progress of sciences and humanities together.'
24. Geddes, *Cities*, p. 186.
25. Kitchen, Paddy, *A Most Unsettling Person: An Introduction to the life and Ideas of Patrick Geddes* (London, Gollancz 1975), p. 149. The quote originates from an unpublished typescript in the Geddes papers at Strathclyde University (SUA, T-GED 5/3/70, p. 4). See Welter, *Arcades*, p. 330, n. 4. The item is an unpublished paper on aesthetics that has been suggested by Welter to date from about 1899.
26. SUA T-GED 9/2175. Letter from Margaret M. Mackintosh to Mrs Geddes written from 6 Florentine Terrace, Hillhead, Glasgow. It has no date but is datable to 1912–14 inclusive, with the earlier date more likely due to the reference to the Duncans' new house (they moved into their new house in St Bernard's Terrace, Edinburgh in 1912). The Mackintoshes left Glasgow in 1914.
27. Welter, Volker, 'Arcades for Lucknow: Patrick Geddes, Charles Rennie Mackintosh and the Reconstruction of the City', *Architectural History*, Vol. 42, 1999, pp. 316–32.
28. The links between Geddes and Mackintosh are further underlined by the fact that it was Geddes who intervened (via his daughter, Norah) on Mackintosh's behalf when, due to the latter's international mail, his interest in drawing and his 'foreign' accent, he was under suspicion of spying in the English county of Suffolk in 1915. See Howarth, Thomas, *Charles Rennie Mackintosh and the Modern Movement* (London: Routledge & Kegan Paul, 1952), p. 196.

29. Geddes, *Cities*, pp. 374–5.
30. For more detail on this project, see Boardman, Philip, *Worlds of Patrick Geddes* (London: Routledge, 1978), pp. 211–13; Meller, Helen, *Patrick Geddes* (London: Routledge, 1990), pp. 147–8.
31. Saint, Andrew, 'Ashbee, Geddes, Lethaby and the Rebuilding of Crosby Hall', *Architectural History*, Vol. 34, 1991, pp. 206–23.
32. Abercrombie, Patrick, *Town and Country Planning* (London: Thornton Butterworth, 1933), p. 129. For Abercrombie, Geddes's unsettling interdisciplinarity was a necessary counter to tempting utopian oversimplifications. In a letter reproduced as an appendix to Amelia Defries's book on Geddes, Abercrombie had written: 'Bluntly, what Geddes taught was, that if you wish to shape the growth of a town, you must study it', and not just the town, but its relation to the region and with the world at large. Defries, Amelia, *The Interpreter: Geddes* (London: Routledge, 1927), pp. 322–5, p. 323.
33. Geddes, *Cities*, p. 258.
34. For more on Lady Aberdeen, see Pentland, Marjorie, *A Bonnie Fechter: The Life of Ishbel Marjoribanks* (London: Batsford, 1952), and *Onwards and Upwards*, edited by James Drummond (Aberdeen: Aberdeen University Press, 1983).
35. Meller, *Geddes*, p. 185.
36. Bowe, Nicola Gordon and Cumming, Elizabeth, *Arts and Crafts in Dublin and Edinburgh* (Dublin: Irish Universities Press, 1998).
37. Geddes, *Cities*, p. 263.
38. Boardman, *Geddes*, pp. 240–3.
39. The main competition was from the German exhibit, but as Geddes commented: 'The Germans may have their frames and labels, but we have the ideas'. Ibid., p. 242.
40. Weirter, Louis, *The Story of Edinburgh Castle* (London: Harrap, 1913).
41. Simpson, Michael, *Thomas Adams and the Modern Planning Movement* (London: Mansell, 1985), pp. 2–3.
42. Gruffudd, Pyrs, 'Back to the Land: Historiography, Rurality and the Nation in Interwar Wales', *Transactions of the Institute of British Geographers*, New Series, Vol. 19, No. 1, 1994, pp. 61–77.
43. An outline of Geddes's planning activities in Dublin can be found in Meller, *Geddes*, pp. 182–9.
44. Simpson, *Adams*, p. 66.
45. Geddes, Patrick, *Cities in Evolution* (London: Williams and Norgate, 1915), p. ix. For an account of the sinking of the *Clan Grant*, see Dan van der Vat, *The Last Corsair: The Story of the Emden* (London: Hodder and Stoughton, 1983), pp. 78–81.

CHAPTER 11   **Ecological Research in Dundee**

1. Geddes, Patrick and Victor Branford, 'Prefatory Note', *The Evergreen: Book of Autumn* (Edinburgh: Patrick Geddes and Colleagues), p. 8.
2. NLS MS 10537, f. 56, 22 September 1904.
3. For insight into the importance of Geddes's links with Flahault, not least with reference to their mutual interest in the value of mural art, see Thomson, Belinda, 'Patrick Geddes's "Clan d'Artistes": some elusive French connections', in *Patrick Geddes: The French Connection*, edited by Frances Fowle and Belinda Thomson (Oxford: White Cockade, 2004), pp. 47–68.
4. Burnett, John H., *The Vegetation of Scotland* (Edinburgh: Oliver and Boyd, 1964), p. 1.
5. Ibid., p.1.
6. Sheail, John, *Seventy-five Years in Ecology: The British Ecological Society* (Oxford: Blackwell, 1987), p. 8.
7. Ibid., pp. 1–2.
8. Mather, Alexander S., 'Geddes, Geography and Ecology: The Golden Age of Vegetation Mapping in Scotland', *Scottish Geographical Magazine*, Vol. 115, No. 1, 1999, pp. 35–52, p. 49.
9. Ibid.
10. Ibid.
11. Maus, Heinz, *A Short History of Sociology* (London: Routledge & Kegan Paul, 1962), p. 47.
12. For an assessment of Herbertson, see Jay, L. J., 'Andrew John Herbertson', in *Geographers: Biobibliographical Studies*, Vol. 3, 1979, pp. 85–92.
13. Livingstone, David N., *The Geographical Tradition: Episodes in the History of a Contested Enterprise* (Oxford: Blackwell, 1992), p. 282. Fleure was (as noted in chapter ten) Geddes's assistant at the cities exhibition in Dublin.
14. Stevenson, Iain, 'Patrick Geddes', in *Geographers: Biobibliographical Studies*, Vol. 2. 1978, pp. 53–65, p. 59: 'the major responsibility for the making of the Geddesian tradition in geography lies firmly with H. J. Fleure'.
15. Sheail, *Ecology*, pp. 6–10.
16. Ibid., p. 9.
17. As with *The Evergreen* and *Carmina Gadelica*, the printer was T. and A. Constable.
18. Smith, Robert, 'Botanical survey of Scotland. I. Edinburgh District', *Scottish Geographical Magazine*, Vol. 16, 1900, pp. 385–415; Smith, Robert, 'Botanical survey of Scotland. II. North Perthshire District', *Scottish Geographical Magazine*, Vol. 16, 1900, pp. 441–67.
19. Geddes, Patrick, 'Robert Smith BSc, University College Dundee', *Scottish Geographical Magazine*, Vol. 16, 1900, pp. 597–9.

20. Smith, William G., 'A botanical survey of Scotland', *Scottish Geographical Magazine*, Vol. 18, pp. 132–9; Smith, William G., 'Botanical survey of Scotland. III and IV. Forfar and Fife', *Scottish Geographical Magazine*, Vol. 20, 1904, pp. 617–28; and Vol. 21, pp. 4–23, 57–83 and 117–26.

21. Geddes, Patrick, 'A Great Geographer: Élisée Reclus', *Scottish Geographical Magazine*, Vol. 21, 1905, pp. 490–6 and pp. 548–55.

22. Robson, B. T., 'Geography and Social Science: The Role of Patrick Geddes', *Geography, Ideology and Social Concern*, edited by D. R. Stoddart (Oxford: Blackwell, 1981), pp. 186–205, p. 187.

23. For an assessment of how Geddes's planning work influenced the development of geography as a discipline in Britain, see Matless, David, 'Regional Surveys and Local Knowledges: The Geographical Imagination in Britain', *Transactions of the Institute of British Geographers*, New Series, Vol. 17, No. 4, 1992, pp. 464–80.

24. My thanks to Ann Prescott for first drawing Geddes's scheme to my attention. For detailed consideration of it, see Arnold, Mary, 'A Dream of a Bonnier Dundee – Geddes's ideas for Parks and Gardens in Dundee', in *The Artist and the Thinker: John Duncan and Patrick Geddes in Dundee*, edited by Matthew Jarron (Dundee: University of Dundee Museum Services, 2004), pp. 42–54.

25. It is an irony that one of Dundee's finest buildings, the Town House by William Adam, was demolished in the year of Geddes's death. Plans by James Thomson to re-erect it, as Geddes had with Crosby Hall, were never put into effect. See Crumley, Jim, 'The Heh Street', *The Scots Magazine*, New Series, Vol. 166, No. 2, February 2007, pp. 198–203. For an assessment of Thomson, see Harris, Bob, 'City of the future: James Thomson's vision of the city beautiful', in *Victorian Dundee: Image and Realities*, edited by L. Miskell, C. A. Whatley and B. Harris (East Linton: Tuckwell, 2000), pp. 169–83.

26. Geddes, Patrick, *Cities in Evolution* (London: Williams and Norgate, 1915), p. 378.

27. For proceedings, abstracts of papers, programmes, membership, etc., see *Report of the Eighty-Second Meeting of the British Association for the Advancement of Science, Dundee 1912* (London: John Murray, 1913). For popular interest in the meeting, see *The Piper o' Dundee*, various issues for August and September 1912, Dundee University Archive Services MS. 88/11/6. For information about Dundee in 1912, see *Handbook and Guide to Dundee and District: Prepared for the Members of the British Association for the Advancement of Science, on the occasion of their visit to Dundee, under the direction of the Local Publications Committee*, edited by A. W. Paton and A. H. Millar (Dundee: David Winter, 1912).

28. D'Arcy Thompson was also a key figure in the local organising committee for the Dundee meeting.

29. *Geographical Journal* XL, July–December 1912 pp. 537–50. (Papers by E. A. Reeves, C. Markham and W. S. Bruce).

30. Spufford, Francis, *I May be Some Time: Ice and the English Imagination* (London: Faber and Faber, 1996), pp. 273–88.

31. Speak, Peter, 'Introduction', in Bruce, William Spiers, *The Log of the Scotia Expedition 1902–1904*, edited by Peter Speak (Edinburgh: Edinburgh University Press, 1992).

32. Spufford, *Some Time*, p. 288.

33. *Geographical Journal*, Vol. 40, 1912, p. 548.

34. Ibid., p. 549.

35. Swinney, Geoffrey N., 'The training of a polar scientist: Patrick Geddes and the student career of William Spiers Bruce', *Archives of Natural History*, Vol. 29, No. 3, 2002, pp. 287–301; Speak, Peter, *William Spiers Bruce: Polar Explorer and Scottish Nationalist* (Edinburgh: National Museums of Scotland, 2003). Bruce's influence on the polar explorer James Wordie can also be noted: see Smith, Michael, *Sir James Wordie, Polar Crusader* (Edinburgh: Birlinn, 2004), pp. 125–31.

36. *Life in the Antarctic* (Glasgow: Gowans and Gray, 1907). This book was part of a series of small-format photographic nature books which included *Our trees and how to know them* (sixty photographs by Charles Kirk), which had a cover by Jessie M. King. Other titles in the same visually intensive small format included a series of affordable books of black-and-white photographic reproductions of the art of the old masters, complete with Celtic revival covers.

37. SUA T-GED 1107.

38. *Illustrated Catalogue of a Loan Collection of Paintings Watercolours and Engravings in the Victoria Galleries Dundee on the Occasion of the British Association Meeting with an introduction by A.H. Millar* (Dundee, 1912).

39. With respect to the continuing iconographical interest of *Riders of the Sidhe*, one can note that the painting had a significant role in the exhibition *Celts: Art and Identity* held at the British Museum, London, in 2016.

40. Millar, A. H., introduction, *Illustrated Catalogue*, p. 8.

41. Ibid., p. 9.

CHAPTER 12 Dramatising the Past and Informing the Future

1. McGegan, Edward, 'Patrick Geddes', *The Scottish Bookman*, Vol. 1, No. 4, 1935, pp. 99–106, p. 104.

2. SUA T-GED 12/3/31.

3. Geddes, Patrick, *The Masque of Learning and its Many Meanings* (Edinburgh: Patrick Geddes and Colleagues, 1912). My attribution of the frontispiece to Cursiter is based on the 'SC' monogram which appears there, tracked to Cursiter through the monogram on his book cover for

F. C. Inglis's *Dear Auld Reekie* published in 1925. Cursiter also contributed to a book of quotations, *A Garden Enclosed*, published in 1911 to raise funds for the Outlook Tower.

4. See Cursiter's decorations for Gunn, John, ed., *The Orkney Book* (Edinburgh: Nelson,1909).

5. For the wider cultural political context, see McArthur, Euan, *Scotland, CEMA and the Arts Council, 1919–1967: Background, Politics and Visual Art Policy* (Farnham: Ashgate, 2013).

6. My information is taken from the report in *The Scotsman* of 11 June 1908. See also Kinchin, Juliet, 'Art and History into Life: Pageantry Revisited in Scotland', *Journal of the Scottish Society for Art History*, No. 2, 1997, pp. 42–51.

7. For example, Riders of the Sidhe, Cuchullin, Ossian and Columba.

8. Kemplay, John, *John Duncan: A Scottish Symbolist* (San Francisco: Pomegranate, 1994), p. 71.

9. As noted in chapter ten, Otto Schlapp was also the artist who in 1912 contributed an image of Edinburgh Castle to Geddes's publication *The Blue Blanket*, used in 1913 as the endpaper design for Louis Weirter's *The Story of Edinburgh Castle*.

10. NLS MS 10509, ff. 111–15a.

11. These 'Riders of the Sidhe' predate Duncan's painting of the same subject by three years.

12. See MacGregor, Mary, n.d., *Stories of King Arthur's Knights told to the Children*, and Parker, Winifred M., *Na Daorie Sidhe in Uirsgeulen Eile – Gaelic Fairy Tales*, 1908. In 1929 Cameron published a remarkable series of images of the Old Town of Edinburgh, including Geddes's renovated Riddle's Court, as illustrations to Grierson, Flora, *Haunting Edinburgh* (London: John Lane, 1929).

13. Howarth, Thomas, *Charles Rennie Mackintosh and the Modern Movement* (London: Routledge & Kegan Paul, 1952), pp. 230–1.

14. Christian, Jessica and Stiller, Charles, *Iona Portrayed* (Inverness: New Iona Press, 2000), pp. 66, 70.

15. Carmichael, Eoghan Kenneth, *The Elements of Celtic Art* (Glasgow: An Comunn Gaidhealach, 1922).

16. For an image, see Cumming, Elizabeth, *Hand, Heart and Soul: the Arts and Crafts Movement in Scotland* (Edinburgh: Birlinn, 2006), p. 181.

17. Campbell, John Francis, *Popular Tales of the West Highlands*, Vol. 4 (Edinburgh: Edmonston and Douglas, 1862), pp. 381–403. For context, see Macdonald, Murdo, 'The Visual Dimension of *Carmina Gadelica*', in *The Life and Legacy of Alexander Carmichael*, edited by Domnhall Uilleam Stiùbhart (Port of Ness: Islands Book Trust, 2008), pp. 135–45, p. 140; Macdonald, Murdo, 'The Visual Preconditions of Celtic Revival Art in Scotland', *Journal of the Scottish Society for Art History*, pp. 16–21, pp. 19–20.

18. Evans, Sebastian, trans., *The High History of the Holy Graal* (London: Dent, 1903).

19. Geddes, *Masque of Learning*, p. 32.

20. Geddes, Patrick, *The Masque of Ancient Learning* (Edinburgh: Patrick Geddes and Colleagues, 1912), printed text inside front cover.

21. Ibid.

22. In the first edition of *The Masque of Ancient Learning* Cursiter is credited (p. x) for his design of a poster. This is presumably the design that appears on the back cover. In the 1912 edition the back cover design includes the text: 'The Edinburgh Masquers Outlook Tower'. In later editions this text is erased, as are the acknowledgements to Edinburgh supporters (including Cursiter) and details of the original Edinburgh performers in the prefatory material pp. x–xxviii. This is presumably to make the text less specific, and thus more appropriate for distribution at the London performances. The final version of the *Masques* text brought ancient, medieval and modern together in Geddes, Patrick, *Dramatisations of History* (Bombay and Karachi: The Modern Publishing Co., 1923). Two other publishers are also given on the title page: London: Sociological Publications, and Edinburgh: Patrick Geddes and Colleagues. The printer is the Huxley Press, Madras.

23. Ibid.

24. Geddes, *Dramatisations*, p. i.

25. Geddes, *Ancient Learning*, p. v.

26. Geddes, *Ancient Learning*, p. vi.

27. Taking this classical insight further: 'In the gods on Olympus [Geddes] discovered the ideal embodiment of the Seven Ages of Man'. Mumford, Lewis, *The Condition of Man* (New York: Harcourt, Brace, 1944), p. 384.

28. I owe this notion of Jung's archetypes as 'methods of thought' to a lecture by Boris Semeonoff at the University of Edinburgh in the late 1970s.

29. Geddes, *Ancient Learning*, p. vi.

30. After the First World War, Geddes would return to this ideal of a Europe united by civic values: 'The reunion of Europe, then, can most strongly, even if slowly, be made through the education of travel. Not merely in the recent tourist spirit, at least in the cruder forms; but in that combining of the best of modern cultural travel with something of the old spirit of pilgrimage which that helps effectively to renew. The Brownings and Ruskin in Italy were examples of this union in their day: why not renew it more widely? As Europeans grow more tolerant and more sympathetic ... our scheme of educational travel will grow and spread into fuller pilgrimages ... throughout Baltic and Mediterranean lands alike, from Scandinavia to Spain, and thence to Greece and beyond. Why not east and west, from Russia and Ireland, indeed to America as well? – with ever increasing appreciation of all their regional and civic interests, the natural, the spiritual, and the temporal together, and in aspects historic,

actual and incipient. Does this seem "Utopian"? It is after all but what the tourist and the wandering nature-lover, the art-student, and the historian have long been doing, and what the regional agriculturalist and the town planner are now in their turn doing. Today it lies with re-education, with reconstruction, and with re-religion as well, to organise these contacts more fully.' Geddes, Patrick, *The Life of Sir J C Bose* (London: Longmans, 1920), p. 118.

31. Geddes, *Ancient Learning*, p. vi.
32. Branford, Victor, *St Columba: A Study of Social Inheritance and Spiritual Development* (Edinburgh: Patrick Geddes and Colleagues, 1912).
33. Branford, Victor, *St Columba: A Study of Social Inheritance and Spiritual Development* (Edinburgh: Patrick Geddes and Colleagues, 1913).
34. Ibid., p. 4.
35. Ibid., p. 5.
36. NLS MS 10513 ff. 136–7. Letter from Patrick Geddes to Stanley Hall, 7 December 1912.
37. Ibid.
38. SUA T-GED 12/3/31. Letter from John Duncan to Patrick Geddes. No date, probably Spring 1912.
39. McDonald, Allan, *Eilein na h-Oige: The Poems of Fr Allan McDonald*, edited by Ronald Black (Glasgow: Mungo, 2002), p. 37.
40. Ibid., p. 87.
41. Warrack, Grace, ed., *From the Isles of the West to Bethlehem* (Oxford: Blackwell, 1921), facing p. 100.
42. McDonald, *Poems*, p. 87.
43. Private collection.
44. Mumford, *Condition of Man*, p. 388.
45. Geddes, *Masque of Learning*, p. 36.
46. Edinburgh: Patrick Geddes and Colleagues, 1897.
47. Macleod, Fiona, *The Divine Adventure: Iona: Studies in Spiritual History*, edited by Mrs William Sharp (London: Heinemann, 1919; Vol. IV of the Uniform Edition, 1910). The essay, under the title *The Isle of Dreams*, also appeared in a shortened form in 1905 as one of the 'Iona Books' published by T. N. Foulis.
48. Cameron's work was prominently covered in *The Studio* and in due course two volumes of his etchings were issued by the magazine in the *Modern Masters of Etching* (London: The Studio, 1925 and 1932).
49. Errington, Lindsay, *Scotland's Pictures* (Edinburgh: National Galleries of Scotland, 1990), p. 70.
50. With reference the Celtic significance of Ben Ledi, it is of interest that the nearest Christian foundation to Ben Ledi is St Bride's Chapel on Loch Lubnaig. A number of views of this mountain by Cameron appear as illustrations to Seton Gordon's *Highways and Byways of the Central Highlands* published in 1948.

51. Geddes, *Masque of Learning*, p. 31.
52. MacDonald, Keith Norman, *In Defence of Ossian* (Oban: Privately Published, 1906).
53. Geddes, *Masque of Learning*, p. 31.
54. Ibid.
55. Geddes also compares this division of activities to Hinduism: 'despite all difference and remoteness, these four [i.e. the riders] will be seen to correspond in principle to the four castes of our first Hindu scene. The correspondence may perhaps be related with those of comparative philology and of folklore: yet it is of even more general, indeed universal range.' Ibid.
56. Ibid., pp. 31–2.
57. Dehaene, Michiel, 'Survey and the assimilation of a modernist narrative in urbanism', *The Journal of Architecture*, Vol. 7, 2002, pp. 33–55.
58. My thanks to Elizabeth Cumming and Alan Crawford for discussion of this point.
59. MacCarthy, Fiona, *The Simple Life: C. R. Ashbee in the Cotswolds* (Berkeley: University of California Press, 1981), pp. 174–6.
60. Ashbee, Charles Robert, 'The "Norman Chapel" Buildings at Broad Campden, in Gloucestershire', *Studio*, Vol. 41, 1907, pp. 289–96, pp. 295–6.
61. Coomaraswamy, Ananda K., *Mediaeval Sinhalese Art* (Broad Campden: Essex House Press, 1908).
62. Harrod, Tanya, 'Ananda Coomaraswamy', *Crafts*, November–December 1996, pp. 20–3; Harrod, Tanya, *The Crafts in Britain in the 20th Century* (New Haven: Yale, 1999), pp. 44–5 and pp. 177–9. One of the design critics she notes is Amelia Defries (*Crafts in Britain*, p. 177).
63. Abercrombie, Patrick, Branford, Victor, Desch, C., Geddes, Patrick, Saleeby, C. W. and Kilburn Scott, E., *The Coal Crisis and the Future: A Study of Social Disorders and their Treatment* (London: Williams and Norgate, 1926).
64. Mairet, Philip, *Pioneer of Sociology: The Life and Letters of Patrick Geddes* (London: Lund Humphries, 1957).
65. In 1932, towards the end of his life, Geddes became president of the New Europe Group of the Adler Society. The formation of this group was a response to the persecution of German members of the Society as a result of the rise of Hitler. See Mairet, *Pioneer of Sociology*, p. 210.
66. Lipsey, Roger, 1977, *Coomaraswamy, His Life and Work* (Princeton: Princeton University Press, 1977), pp. 45–6. It has been suggested that Nivedita had left India for a period under threat of prison or deportation for her nationalist activities. For qualification of this, see Guha-Thakurta, Tapati, *The Making of a New 'Indian' Art* (Cambridge: Cambridge University Press, 1992), pp. 171–2.
67. Sharp's Anglo-Celticism has something in common with the strongly

westernised Hindu imagery of the painter Ravi Varma. Both are paradoxi-cal figures. Geeta Kapur has described Varma as both a 'prime aspirant in the Indian artists' passage to the modern' and at the same time an 'obvious anachronism': Kapur, Geeta, 'Ravi Varma: Historicizing Representation', in *Indian Painting: Essays in Honour of Karl J. Khandalavala*, edited by B. N. Goswamy (New Delhi: Lalit Kala Akademi, 1995), pp. 223–43, p. 223.

68. Carmichael's interest in the links between Indian and Celtic cultures is expressed in Carmichael, Alexander, *Carmina Gadelica* (Edinburgh: Norman Macleod, 1900), pp. xxix–xxx.

69. Geddes is quoted on p. 107 in the essay 'Education in India'. Carmichael is quoted in an extensive footnote on p. 111 in the essay 'Memory in India'. The footnote draws a parallel between the high quality of language preserved in the oral tradition of rural working people of seventeenth-century Ceylon and the Gaelic speakers of the Scottish Highlands.

70. *Quarterly Journal of the Mythic Society*, 37, 1947, pp. 1–14. Reprinted in Coomaraswamy, Ananda, *What is Civilization? and other Essays* (Ipswich: Golgonooza, 1989), pp. 94–106.

71. Coomaraswamy, Ananda, 'The Bugbear of Literacy', in *Am I my Brother's Keeper?* (New York: Day, n.d., c. 1947) p. 24 and pp. 33–4.

72. Bain, George, *The Methods of Construction of Celtic Art* (Glasgow: Maclellan, 1951), p. 20.

73. Ibid., p. 16.

74. Coomaraswamy, Ananda, *Art and Swadeshi* (Madras: Ganesh, 1912), p. 135.

75. Sadler was later to describe Geddes's ideas about university organisation as 'pure gold'. See Kitchen, *Geddes*, p. 268. Sadler was also an early advocate of the painting of Gauguin. Indeed, a high point of that artist's Breton period, *Vision after the Sermon*, now in the National Gallery of Scotland, was previously in Sadler's collection.

76. NLS MS 10543 f. 52.

77. Geddes, *Ancient Learning*, p. 77.

78. NLS MS 10543 f. 52. Letter dated 30 March 2012.

79. NLS MS 10543 f.10. Letter dated 15 January 1912.

80. Mitter, Partha, *Art and Nationalism in Colonial India* (Cambridge: Cambridge University Press, 1994), p. 312.

81. *The Burlington Magazine* of 1913 unites the writing of younger-generation figures inspired by Arts and Crafts ideas, such as Coomaraswamy and A. P. Laurie, alongside that of an older generation of Arts and Crafts activists such as W. R. Lethaby and Baldwin Brown.

82. SUA T-GED 5/1/20.

83. 'Kawaguchi' is wrongly transcribed in the typescript as, among other things, 'Kawagudu'.

84. Geddes, *Ancient Learning*, p. 75.

85. Mitter, *Art and Nationalism*, p. 312.
86. The idea became a reality in Geddes's final project, the Collège des Ecossais at Montpellier. An illuminating account was written by Geddes's old Dundee friend and supporter, Robert C. Buist. In his first paragraph Buist notes the presence of Scots Colleges throughout Europe prior to the Reformation, emphasising that Geddes's establishment of the Scots College in Montpellier was of a piece with all his other cultural revivalist activities. Buist reminds us that 'we have almost forgotten that a college was a society of students rather than an engine for teaching': Buist, R. C., 'The Newest Scots College. Some Account of the Work of Patrick Geddes', *The Scots Magazine*, New Series, Vol. VIII, No. 5, February 1928, pp. 321–4. See also Andrieu, Jean-Paul and Geddes, Marion (eds.), *La Colline et le Monde: Patrick Geddes et le Collège des Ecossais, Montpellier, France, vus par ses contemporains* (Editions de l'Espérou: Montpellier, 2019); and Chabard, Pierre, 'Paris-Montpellier-Domme: French Migrations of the Outlook Tower (1900–1924–1937)', *Journal of the Scottish Society of Art History*, Vol. 9, 2004, pp. 79–86.
87. Geddes, *The Masque of Medieval and Modern Learning* (Edinburgh: Patrick Geddes and Colleagues, 1912), pp. 13–14.
88. Ibid., p. 14.
89. Ibid.
90. Ibid., p. 11.
91. Ibid.
92. Ibid., pp. 11–12.
93. Ibid., p. 46.
94. Ibid., p. 47.
95. Ferguson, Adam, *An Essay on the History of Civil Society* (1767), edited by Duncan Forbes (Edinburgh: Edinburgh University Press, 1966). See also Allan, David, *Adam Ferguson* (Edinburgh: Edinburgh University Press, 2006), pp. 136–9.
96. Geddes, *Medieval and Modern*, p. 47.
97. SUA T-GED 12/3/31.
98. Printed by J. C. Thomson at The Mercat Press in Edinburgh.
99. Chrystal, George, trans., *The Meditations of Marcus Aurelius Antoninus* (Edinburgh: Otto Schulze, 1902).
100. Elfick, Ian and Harris, Paul, *T N Foulis: The History and Bibliography of an Edinburgh Publishing House* (London: Werner Shaw, 1998), p. 9.
101. MacFall, Haldane, 'Joseph Simpson: Caricaturist', *The Studio*, Vol. 35, 1905, pp. 20–5. In due course Simpson published drawings in *Rhythm* (July and September 1912), of which J. D. Fergusson was the editor. Haldane MacFall also published in *Rhythm*, and also wrote about Fergusson for *The Studio*.
102. Leask, William Keith, 'The Foulis Press 1905–1913', *The Thistle*, July 1913, pp. 115–16, p. 116.

103. *The Blue Blanket: An Edinburgh Civic Review* (Edinburgh: T. N. Foulis, 1912); cf. Pennecuik, Andrew, *The History of the Blue Blanket or Crafts-Man's Banner* (Edinburgh: David Bower, 1722).

104. Home was a fine visual recorder of Edinburgh and one of the moving spirits behind the formation of the Old Edinburgh Club.

105. Marion Newbigin was a notable geographer and editor of the *Scottish Geographical Magazine*.

106. Traquair, Ramsay and Mears, Frank, 'Public Monuments', *The Blue Blanket*, No. 1, 1912, pp. 68–80. In the same year both men acted in *The Masque of Medieval and Modern Learning*.

107. Geddes and Foulis shared membership of the Scottish Arts Club. For Geddes's link, see Reiach, Alan et al., *The Scottish Arts Club 1874–1974* (Edinburgh: Scottish Arts Club, 1974), p. 18. For Foulis's link, see Elfick, Ian and Harris, Paul, *T N Foulis: The History and Bibliography of an Edinburgh Publishing House* (London: Werner Shaw, 1998), p. 12.

108. A head of Krishna from Ananda Coomaraswamy's own collection, since 1918 in the collection of The Metropolitan Museum of Art, New York, seems to have served as a model for John Duncan in one of his illustrations to Donald A. Mackenzie's *Myths of Crete and Pre-Hellenic Europe*. It is reproduced as the frontispiece to Coomaraswamy's influential introductory work *The Arts and Crafts of India and Ceylon*. Duncan's debt to Minoan wall paintings is also clear (as one would expect) but the stylistic and colouristic comparisons between Duncan's *Ladies of the Minoan Court* and Coomaraswamy's frontispiece are marked. It is likely that Duncan saw this work in Coomaraswamy's book, but one can note that it was also reproduced as part of a loose-leaf advertising insert for T. N. Foulis's publications. This insert appeared in *The Studio* around 1913. It is reproduced in Elfick and Harris, *T N Foulis* as a colour insert bound between pp. 190 and 191.

109. Antliff, Allan, *Anarchist Modernism* (Chicago: University of Chicago Press, 2001), pp. 132–3; Crouch, James S., *A Bibliography of Ananda Kentish Coomaraswamy* (New Dehli: Manohar, 2002), p. 109.

110. Almost fifteen years earlier, in his first book *Mediaeval Sinhalese Art* (Broad Campden, Essex House Press, 1908), p. viii, Coomaraswamy had noted Penty's *Restoration of the Guild System* (London: Swan Sonnenschein, 1906).

111. Penty, Arthur J., *Post-Industrialism* (New York: Macmillan, 1922), p. 14: '. . . the term Post-Industrialism . . . I owe to Dr A. K. Coomaraswamy'.

112. Published in 1910.

113. Laurie, Arthur Pillans, ed., *The Teacher's Encyclopaedia: Theory, Method, Practice, History and Development of Education at Home and Abroad* (London: Caxton, 1912). The section on Comenius is by Alexander Darroch and can be found in Vol. 7, pp. 133–8.

114. Jack Yeats also published in *Rhythm* in November 1912.

115. Geddes, Patrick, *Dramatisations of History* (Bombay and Karachi: Modern Publishing, 1923), p. iv. This publication brought together the *Masque of Ancient Learning* and the *Masque of Medieval and Modern Learning* in one volume.

CHAPTER 13   Looking East

1. Khursheed, Anjam, *The Seven Candles of Unity: The Story of Abdul-Baha in Edinburgh* (London: Bahai Publishing Trust, 1991), p. 47.
2. Barbour, George Freeland, *The Life of Alexander Whyte* (London: Hodder and Stoughton, 1923), pp. 554–5.
3. The larger part of the Free Church had joined with the United Presbyterian Church to form the new body in 1900. For a convenient diagram of the schisms and reunifications of the Presbyterian churches, see Haldane, Elizabeth, *Scotland of Our Fathers: A Study of Scottish Life in the Nineteenth Century* (London: Maclehose, 1933), pp. 132–3.
4. Khursheed, *Seven Candles* (London: Bahai Publishing Trust, 1991), p. 70. Khursheed reproduces a page advertising the meetings from *The Scotsman* newspaper of Monday 6 January 1913. The first meeting (chaired by John Kelman) was at the Freemasons' Hall on Tuesday 7 January and was organised by the Esperanto Society. The New College meeting chaired by Geddes took place on Wednesday 8 January and was arranged by the Outlook Tower Committee. On Thursday 9 January, Abdul Baha gave an address at the Theosophical Society at its premises in Great King Street.
5. Khursheed, *Seven Candles*, p. 117.
6. Senior, Alan, 'Editorial', *Circles: The Magazine of the Theosophical Society in Scotland*, No. 21, Spring 1995, p. 2.
7. In the drawing of the same composition published in *The Blue Blanket* in 1912, Duncan uses a tree of life image instead of the swastika.
8. Lankester, Ray, *Secrets of Earth and Sea* (London: Methuen, 1920), p. 191. The multi-cultural spiritual view of the swastika was reprised in 1926 by the Scottish folklorist Donald A. Mackenzie, who wrote that no symbol 'has of recent years aroused more interest among students of antiquities in both hemispheres': Mackenzie, Donald A., *The Migration of Symbols and their Relations to Beliefs and Customs* (London: Kegan Paul, Trench, Trubner and Co., 1926,), p. 1. This interest unfortunately left the swastika open to political misappropriation.
9. Senior, *Circles*, p. 2.
10. The first edition of *The Immortal Hour* was published by T. N. Foulis in 1908. The printer was T. and A. Constable.
11. Benham, Patrick, *The Avalonians* (Glastonbury: Gothick Image, 1993).
12. In his Fiona Macleod persona, William Sharp dedicated his book *The Winged Destiny* (London: Chapman and Hall, 1904) to Dr Goodchild.

See Sharp, Elizabeth, *William Sharp: A Memoir* (London: Heinemann, 1910), p. 381.

13. Benham, *Avalonians*, p. 113.
14. My thanks to Annie V. F. Storr for allowing me to mention her discovery of this work.
15. NLS 10510 f.89.
16. Macleod, Fiona, 'Mary of the Gael', *The Evergreen: Book of Autumn* (Edinburgh: Patrick Geddes and Colleagues, 1895), pp. 123–47.
17. *The Studio*, Vol. 80, 1920, p. 139.
18. Ross, Anne, *Druids* (Stroud: Tempus, 1999), p. 160.
19. *The Studio*, Vol. 70, 1917, p. 93.
20. *The Studio*, Vol. 71, 1917, p. 59.
21. The last mentioned appears as an illustration of an assessment of Duncan's work by E. A. Taylor ('Some Pictures by John Duncan, A.R.S.A.', *The Studio*, Vol. 80, 1920, pp. 139–47). It appears on p. 139. Two *Ossian*-related works are also illustrated – heads of Ossian and Fionn. Taylor was another Scottish contact for Duncan on visits to Paris. He was a notable Glasgow painter and husband of the outstanding illustrator Jessie M. King. In his article on Duncan, Taylor takes pains to note Duncan's links with Patrick Geddes in the 1890s. Taylor also mentions Duncan's 'Celtic group in a pageant in Edinburgh', which must be a reference to the 1908 pageant. He also emphasises the importance of Alexander Carmichael's translations of the Deirdire legend for Duncan.
22. The reproduction of the drawing of *The Riders of the Sidhe* is sometimes loose, sometimes bound between pp. 356 and 357 of *The Blue Blanket*. Page 357 includes a description of the image by John Duncan.
23. Ireland, Nicola, ' "A man of sensitive culture": A Survey of Twelve Albums by John Duncan in the Royal Scottish Academy Collection', *Journal of the Scottish Society for Art History*, Vol. 11, 2006, pp. 52–8, p. 54.
24. *The Studio*, Vol. 65, 1915, p. 104.
25. Mumford saw Yin and Yang as characteristic of Geddes's attitude: 'Implicitly [Geddes] believed in the Chinese principles of Yin and Yang: the alternation of the passive and the active, the internal and the external, the introvert and the extravert moods, was for him the very rhythm of life itself: the key, not alone to sexual differentiation, but to all activities. To block that process was to thwart life: to participate in it was to be one with life's drift and meaning.' Mumford, *Condition of Man* (New York: Harcourt, Brace, 1944), p. 388.
26. *The Scotsman*, 10 June 1930, p. 12. My thanks to Nicola Ireland of the Royal Scottish Academy for drawing my attention to this report of Duncan's work. The Ananda College was founded by the prominent Theosophist Colonel Olcott in 1886; its first head was the Theosophist C. W. Leadbetter. By 1930 it was a site of Buddhist educational resistance to British colonial rule. For more on these works by Duncan, and a valu-

able overview of Theosophy with respect to the Scottish Celtic revival, see Michael Shaw, 'Theosophy in Scotland: Oriental occultism and national identity', in *The Occult Imagination in Britain, 1874–1947*, edited by Christine Ferguson and Andrew Radford (London: Routledge, 2018), pp. 23–40.

27. Defries, Amelia, *Pioneers of Science* (London: Routledge, 1928).
28. For an exploration of these Scottish–Finnish resonances, see Macdonald, Murdo, '*Ossian*, Scottish Narrative and *Kalevala*', European Revivals website (under construction), Ateneum Museum (Finnish National Gallery), Helsinki.
29. John Duncan, notebook dated 1910–11. NLS ACC 6866/3.
30. SUA T-GED 5/3/70, p. 4.
31. Fergusson, John Duncan, *Modern Scottish Painting* (Glasgow: William Maclellan, 1943), pp. 84–8.
32. MacDiarmid, Hugh, *In Memoriam James Joyce* (Glasgow: William Maclellan, 1955).
33. Benham, *Avalonians*, p. 176.
34. Morris, Margaret, *The Notation of Movement* (London: Kegan Paul, Trench, Trubner, 1928). The inscription reads 'To dear Margaret & "Toshie" with much love from Meg, Sep 1928'. Lou Rosenburg's collection.
35. Kenny Munro drew my attention to this via reference to Morris on p. 7 of the prospectus for the Scots College for the session of 1931–2. Further insight was provided by Claire Button of the University of Edinburgh who drew my attention to a flyer for the summer school in the Margaret Morris archive at the Fergusson Gallery in Perth. It includes the passage: 'to be opened in September as a Centre of Culture (artistic and physical) and for Complete Education (Kindergarten to College) in co-operation with Margaret Morris and Professor Patrick Geddes'. Richard Emerson has noted Geddes's assistant Amelia Defries as part of Morris's circle: Emerson, Richard, *Rhythm and Colour* (Edinburgh: Golden Hare, 2018), p. 127.
36. Macdonald, Murdo, 'Rediscovering Nivedita as an Advocate of Visual Art', in *Nivedita: In Search of the Lesser-Known*, edited by Shilpi Das and Santanu Banerjee, Nivedita Project (forthcoming); Macdonald, Murdo, 'Education, Visual Art and Cultural Revival: Tagore, Geddes, Nivedita and Coomaraswamy', *Gitanjali and Beyond*, 1, 2016, pp. 39–57; Macdonald, Murdo, 'Nivedita from a Scottish Point of View', *Prabuddha Bharata/Awakened India*, January 2017, pp. 161–7.
37. Geddes, Patrick, *The Masque of Ancient Learning* (Edinburgh; Patrick Geddes and Colleagues, 1912), p. 75.
38. The illustrations in MacKenzie's book range from Indian bronzes to orientalist images by Warwick Goble. Along with these are images by Abanindranath Tagore's students Nanda Lal Bose and Surendra Nath

Gangoly. In 1918 the former contributed along with Abanindranath Tagore and others to the illustrated Macmillan edition of Rabindranath Tagore's *Gitanjali* and *Fruit-Gathering*. The 'song offerings' of *Gitanjali* were originally published by the India Society in 1912 and introduced by W. B. Yeats. Coomaraswamy's *Buddha and the Gospel of Buddhism* should also be noted here. It was published in 1916, again by Harrap, with illustrations in colour by Abanindranath Tagore and Nanda Lal Bose.

39. Meller, Helen, *Geddes. Patrick Geddes: Social Evolutionist and City Planner* (London: Routledge, 1990), p. 201.

40. Stamp, Gavin, 'The London Years', in *Charles Rennie Mackintosh*, edited by Wendy Kaplan (New York: Abbeville, 1996), p. 207.

41. Welter, Volker, 'Arcades for Lucknow: Patrick Geddes, Charles Rennie Mackintosh and the Reconstruction of the City', *Architectural History*, Vol. 42, 1999, pp. 316–32.

42. Crawford, Alan, 'Lost and Found: Architectural Projects after Glasgow', in *C. R. Mackintosh: The Chelsea Years* (Glasgow: Hunterian Gallery, 1994), p. 5.

43. Lethaby, W. R., *Architecture, Mysticism and Myth* (London: Percival, 1892), p. 149. The passage is from *Fingal*.

44. The interest of this Scottish tradition of analysis of myth is again indicated by Freud's *Totem und Tabu* published in 1913. The four authors most cited by Freud are William Robertson Smith, James Frazer and Andrew Lang, along with the German psychologist Wilhelm Wundt.

45. Fergusson, James, *History of Indian and Eastern Architecture, forming the third volume of the new edition of the History of Architecture* (London: John Murray, 1876), pp. 92–9. See also more detailed treatment in Fergusson, James, *Tree and Serpent Worship or Illustrations of Mythology and Art in India from the Topes of Sanchi and Amravati* (London: India Museum, 1868).

46. Niggl, Reto, *Eckart Muthesius: Indien/India 1930–1939* (Berlin: Goethe-Institutes, 1999), pp. 272–8 and pp. 34–5. My thanks to Richard Carr for drawing this work to my attention.

47. Meller, *Geddes*, p. 204.

48. Geddes, *Indore Report*, part 2, section 1, pp. 1–73.

49. Niggl, *Muthesius*, p. 27.

50. Meller, *Geddes*, p. 236.

51. Ibid., p. 202.

52. NLS MS 19284 (9–18) ff. 13–16, contains an informative eight-page prospectus, *The Darjeeling Summer Meeting, 1917*. See also Boardman, Philip, *The Worlds of Patrick Geddes* (London: Routledge, 1978), pp. 331–2, quoted by Fraser, Bashabi, *The Geddes–Tagore Letters* (Edinburgh: Word Power Books, 2005), p. 29.

53. For example, as noted in chapters two and four respectively, her memoir

of her father, NLS MS 10508, and her clarification of the relationship between her father and D'Arcy Thomson in the dedication of her book of poems, *Intimations and Avowals* (Edinburgh: Moray Press,1944).

54. See, for example, Geddes, Arthur, 'The "Outer" Hebrides', *The New Naturalist*, Summer 1948, pp. 72–6; Geddes, Arthur and Spaven, Frank, *The Highlands and Isles: Their Regional Planning* (Edinburgh: Outlook Tower, 1949); Geddes, Arthur, *The Isle of Lewis and Harris: A Study in British Community* (Edinburgh: Edinburgh University Press, 1955). For an assessment of Arthur Geddes as a geographer, see Andrew T. A. Learmonth, 'Arthur Geddes', in *Geographers: Biobibliographical Studies*, Vol. 2 (London: Mansell, 1978), pp. 45–51.

55. See *Fourteen Songs by Rabindranath Tagore learned, translated and introduced at the Bard's wish by Arthur Geddes*, edited by Marion and Claire Geddes (Bideford: Resurgence Trust, 2011).

56. Geddes, Arthur, *Presenting Tagore in Sound and Sight* (Edinburgh: Adam House, 1961).

57. Personal communication, quoted by Macdonald, Murdo, 'Patrick Geddes: Environment and Culture', in *Think Global, Act Local: the Life and Legacy of Patrick Geddes*, edited by Walter Stephen (Edinburgh: Luath, 2004), p. 63.

58. Geddes, Arthur, *Songs of Craig and Ben*, Vol. 1 (Edinburgh: Serif Books, 1951); Geddes, Arthur, *Songs of Craig and Ben*, Vol. 2 (Glasgow: Maclellan, 1961).

59. Fraser, Bashabi, ed., *A Meeting of Two Minds: Geddes–Tagore Letters* (Edinburgh: Word Power, 2005). See also Fraser, Bashabi, Mukherjee, Tapati and Sen, Amrit, eds., *A Confluence of Minds: The Rabindranath Tagore and Patrick Geddes Reader on Education and Environment* (Edinburgh: Luath, 2018).

60. NLS MS 10544 f. 178. Letter dated 6 December 1914.

61. NLS MS 10545 f. 140. Letter dated 24 May 1917.

62. Campbell, Joseph, *The Hero with a Thousand Faces* (Princeton: Bollingen, 1949), p. 382. Campbell's influence on the film-maker George Lucas can be noted. See Campbell, Joseph, *The Hero's Journey*, edited by Phil Cousineau (New York: HarperCollins, 1990), p. 181.

63. MacArthur, Bessie J. B., 'Memories of Donnachaidh', *The Scots Magazine*, December 1967. Reprinted in *Circles: Magazine of the Theosophical Society in Scotland*, Spring 1995, No. 21, p. 22.

64. Campbell, John Lorne, *Fr Allan McDonald of Eriskay, 1859–1905. Priest, Poet, and Folklorist* (Edinburgh: Oliver and Boyd, 1954), p. 18. Reprinted in *Eilein na h-Òige: The Poems of Fr Allan McDonald*, edited by Ronald Black (Glasgow: Mungo, 2002), pp. 63–74, p. 69.

65. Murray, Amy, *Father Allan's Island* (Edinburgh: Moray Press, 1936), p. 195.

66. Ibid., p. 110.

67. Ibid., p. 108.
68. Jenison, Madge, *Sunwise Turn: A Human Comedy in Bookselling* (New York: Dutton, 1923), pp. 18–19. 'Deasal' is the Gaelic word for 'sunwise'.
69. Antliff, Allan, *Anarchist Modernism* (Chicago: University of Chicago Press, 2001), p. 126.
70. Jenison, *Sunwise Turn*, p. 46
71. Ibid., p. 126.
72. Rilke, Rainer Maria, *Auguste Rodin*, translated by Jessie Lemont and Hans Trausil (New York: Sunwise Turn, 1919).
73. Jenison, *Sunwise Turn*, p. 41.

CHAPTER 14   A Farewell Lecture

1. The lecture marked Geddes's academic moving on from Scotland, first to develop his role as professor of sociology and civics at the University of Bombay, and then to establish a new base, the Collège des Ecossais at Montpellier in the south of France. This last phase of Geddes's activities up to his death in 1932 also included remarkable work in Palestine. See Purves, Graeme, 'A Vision of Zion', The Planning of the Hebrew University of Jerusalem, *The Scottish Review*, No. 21, Spring 2000; Hysler-Rubin, Noah, *Patrick Geddes and Town Planning: A Critical View* (London: Routledge, 2011); Dolev, Diana, *Planning and Building of the Hebrew University, 1919–1948: Facing the Temple Mount* (Lanham: Lexington books, 2016).
2. The conclusion of this era of visual thinking at University College Dundee is symbolised by Geddes writing his classic on the growth and form of cities, *Cities in Evolution*, and D'Arcy Wentworth Thompson writing his classic of biological morphology, *On Growth and Form*. The first edition of *Cities in Evolution* was published in 1915. The first edition of *On Growth and Form* was published in 1917.
3. Defries, Amelia, *The Interpreter: Geddes, the Man and his Gospel* (London: Routledge, 1927), pp. 172–90. The chapter includes both transcribed text and the context of Geddes's lecture. The significance of the lecture was recognised when Defries's chapter was reproduced as Appendix 2 of the revised edition of *Cities in Evolution* (London: Williams and Norgate, 1949). Defries gives no date for the lecture but Geddes's correspondence indicates that he was at University College Dundee in May and June 1919 (NLS MS 10616). June seems to be the correct month, and fits with Defries's reference to 'the long mid-summer evening' at the end of her chapter.
4. Davie, George, *The Democratic Intellect* (Edinburgh: Edinburgh University Press, 1961), p. 24.
5. Ibid., pp. 337–8.
6. Defries, *Geddes*, pp. 173–4.

7. Ibid., p. 175.
8. Ibid.
9. Ibid., p. 178.
10. Ibid., p. 179.
11. Ibid., p. 172. The plants were arranged in the lecture theatre by Arthur Geddes and Defries herself.
12. Ibid., p. 176.
13. Ibid., p. 182.
14. Ibid., p. 184.
15. Ibid.
16. Ibid., p. 186.
17. Ibid.
18. Ibid.
19. Ibid., p. 187.
20. Ibid.
21. Ibid., pp. 187–8.
22. Ibid., p. 188. It is not generally appreciated that Geddes's ideas had contributed to the development of therapies for wartime post-traumatic stress disorder, or as it was called then 'shell-shock'. A key figure in this development was a member of Geddes's Outlook Tower committee, the psychiatrist Arthur Brock, who along with W. H. R. Rivers pioneered the treatment of affected soldiers. One of Brock's patients was Wilfred Owen. The first exercise Brock suggested to Owen was to write an essay about the Outlook Tower, and in July 1917 Owen read it to Brock: 'I perceived that this Tower was a symbol: an Allegory, not a structure but a poetic form. I had supposed it to be a museum and found it [a] philosophical poem: when I had stood within its walls for an hour I became aware of a soul, and the continuity of its idea from room to room, and from storey to storey was an epic.' Quoted in Hibberd, Dominic, *Wilfred Owen: The last Year* (London: Constable, 1992), p. 25. Geddes's notion of the inlook potential of the tower could hardly be better put. My thanks to Lorna J. Waite for making me aware of the Owen–Brock–Geddes link.
23. Ibid.
24. Davie, George, *The Crisis of the Democratic Intellect* (Edinburgh: Polygon, 1986), p. 228.
25. Defries, *Geddes*, p. 188.
26. Ibid., pp. 189–90.
27. Ibid., p. 172.
28. Geddes, Patrick, *The Life and Work of Sir Jagadis Chandra Bose* (London: Longmans, 1920), p. 118.
29. De Carlo, Giancarlo, interview with Peter Wilson, *Newsletter No. 1 of the Edinburgh City of Architecture Bid* (Edinburgh: City of Edinburgh District Council, 1994). Giancarlo De Carlo was in Edinburgh in April 1994 for the ILAUD symposium which he led at Edinburgh College of Art. Papers

were published in *Spazio e Società*, October–December 1994. Six years earlier, De Carlo had proposed an outlook tower for Siena: see Zucchi, Benedict, *Giancarlo De Carlo* (Oxford: Butterworth-Heinemann, 1992), chapter fourteen, 'Outlook Tower, Siena, 1988–1989'; for the influence of Geddes on De Carlo, see Zucchi, chapter two, 'Architecture: The Social Art'. See also 'Trees and Towers', in *Giancarlo De Carlo: Inspiration and Process in Architecture* (Milan: Moleskine, 2011), pp. 24–34.

Appendix 1

Patrick Geddes: A Visual Primer in Ten Images

Figure 1 *Lapis Philosophorum* (philosopher's stone). Geddes's visual signature (with John Duncan's art work) for *The New Evergreen*, 1894, and for *The Evergreen: Book of Winter*, 1896–7. [Murdo Macdonald's collection.]

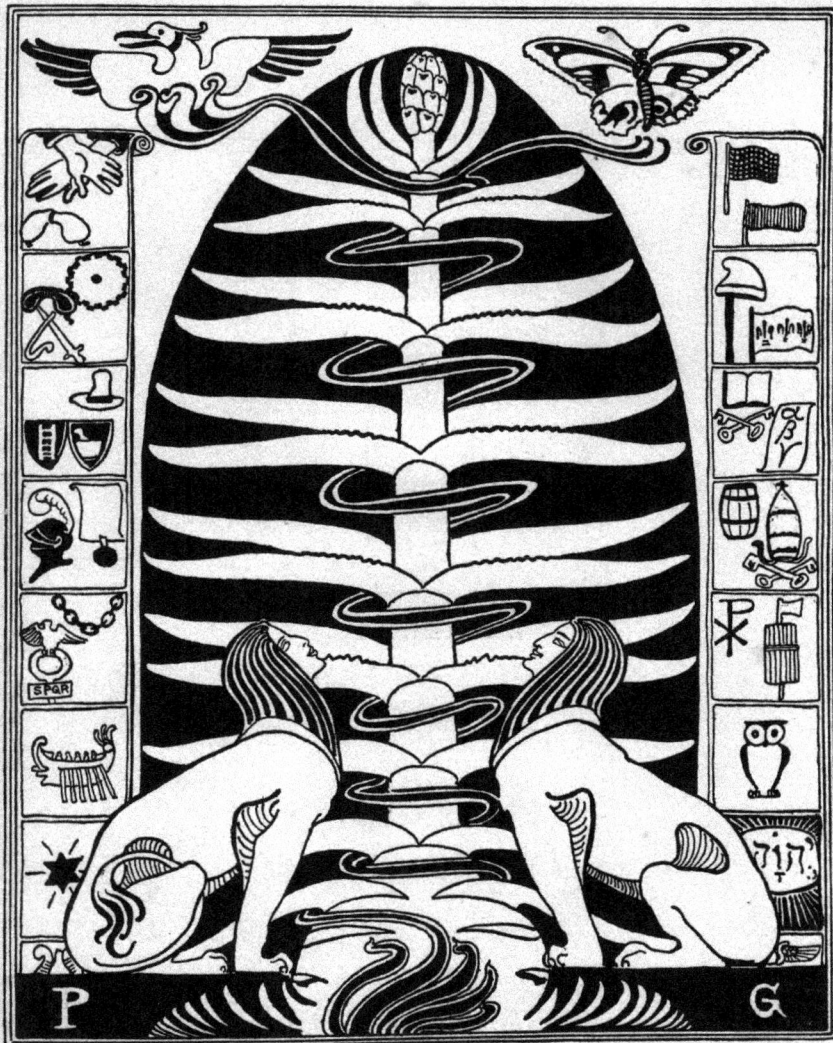

ARBOR SAECULORUM

Figure 2 *Arbor Saeculorum* (tree of the generations). Geddes's visual signature (with John Duncan's art work) for *The Evergreen: Book of Spring*, 1895. [Murdo Macdonald's collection.]

Figure 3 Ramsay Garden sundial with quotes from Robert Burns and Aeschylus. The Aeschylus can be translated as 'time heals all things in aging them'. [Photograph by Murdo Macdonald.]

Figure 4 Lettering and Interlace by Helen Hay, Helen Baxter and Marion Mason surrounding *The Awakening of Cuchullin* by John Duncan in the common room of Ramsay Lodge, 1896. From *Interpretation of The Pictures in the Common Room of Ramsay Lodge* (1944). [Murdo Macdonald's collection.]

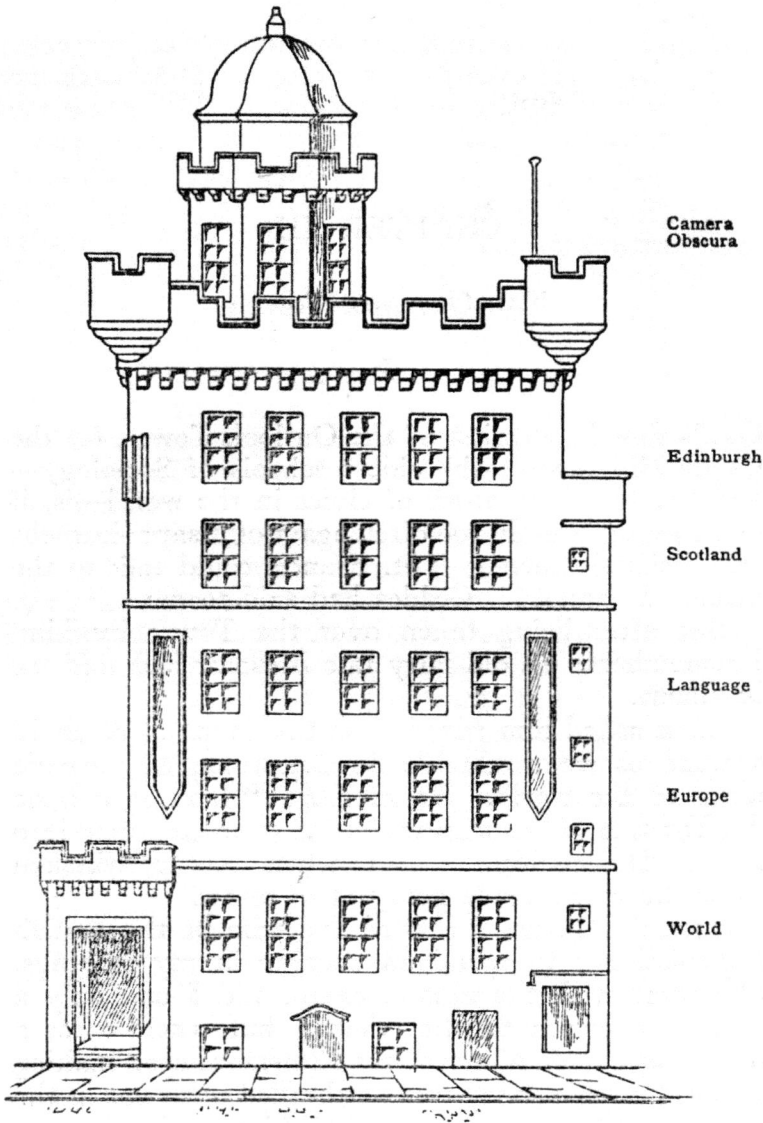

Camera
Obscura

Edinburgh

Scotland

Language

Europe

World

The Outlook Tower, Edinburgh, in diagrammatic elevation, with indication of uses of its storeys—as Observatory, Summer School, etc., of Regional and Civic Surveys, with their widening relation, and with corresponding practical initiatives (from *Cities in Evolution*, by P. Geddes. Williams and Norgate, Publishers).

Figure 5 Diagram of the Outlook Tower from Amelia Defries's *The Interpreter: Geddes* (1927). This image reproduces one published in Geddes's *Cities in Evolution* (1915), and is closely related to one that appears in *A First Visit to the Outlook Tower* (1906). [Murdo Macdonald's collection.]

Figure 6 The stained-glass version of Geddes's valley section diagram described in *A First Visit to the Outlook Tower* (1906). In the care of the Sir Patrick Geddes Memorial Trust. Now located at Riddle's Court, Edinburgh. [Photograph by Murdo Macdonald.]

Figure 7 A sketch of the valley section diagram made by Geddes during his time at University College Dundee. [University of Dundee Archive Services MS 197/9.]

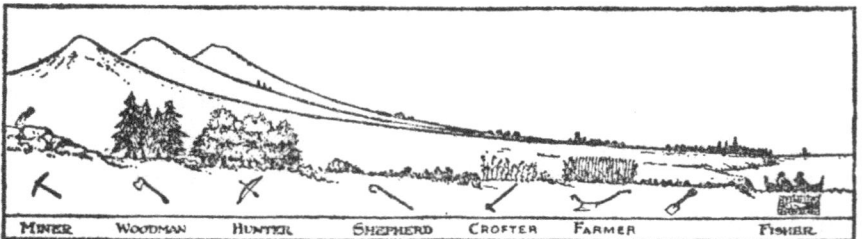

Figure 8 A version of the valley section as it appears in *The Interpreter: Geddes* (1927) by Amelia Defries. [Murdo Macdonald's collection.]

Figure 9 The crest and motto of Geddes's University Hall as it appears in the publication celebrating twenty-five years of University Hall, *The Masque of Learning* (1912). [Murdo Macdonald's collection.]

Figure 10 Geddes's three-dove symbol, standing for sympathy, synthesis and synergy, as it appears on *The Evergreen* and the books of the Celtic Library in the mid-1890s. [Murdo Macdonald's collection.]

Appendix 2

List of Murdo Macdonald's Writings relating to Patrick Geddes and his Milieu

1. 'A Network of Outlooks on Patrick Geddes', *Edinburgh Review*, Issue 88, 1992, pp. 5–9.
2. 'Patrick Geddes – educator, ecologist, visual thinker', *Edinburgh Review*, Issue 88, 1992, pp. 113–19.
3. 'Patrick Geddes: Visual Thinker', in *A Window on Europe: The Lothian in Europe Lectures 1992*, edited by Geraldine Prince (Edinburgh: Canongate, 1993), pp. 229–55.
4. 'Geddes, Patrick', entry for *The New Companion to Scottish Culture*, edited by David Daiches (Edinburgh: Polygon, 1993), pp. 127–8.
5. 'Patrick Geddes e l'intelletto democratico/Art and the Context in Patrick Geddes's Work', *Spazio e Società*, October/December 1994, pp. 28–45.
6. 'Place, Work, Folk: Patrick Geddes and Edinburgh', *Rassegna, 'Themes in Architecture: Edinburgh'*, Vol. 64, No. 4, 1995, pp. 48–51. [Italian edition has the title 'Patrick Geddes e la Outlook Tower: un osservatorio per Edimburgo'.]
7. 'The Visual Thinker: Patrick Geddes', *Edinburgh-Yamaguchi '95 Papers* (Yamaguchi: Yamaguchi Institute of Contemporary Art, 1996), pp. 50–60.
8. 'Weighing the City', in *Brilliant Cacophony: Connecting the Contemporary of Art and Architecture in Edinburgh's Royal Mile*, edited by Lise Bratton (Edinburgh: Scottish Sculpture Trust, 1998), pp. 21–7.
9. 'Sir Patrick Geddes: Pilgrimage and Place', in *In Search of Heritage as Pilgrim or Tourist*, edited by Jan Magnus Fladmark (Shaftesbury: Donhead, 1998), pp. 431–44.
10. 'Art, Ideas and Interdisciplinarity in Scotland', in *Scotland's Art*,

edited by Ian O'Riordan and David Patterson (Edinburgh: City Art Centre, 1999), pp. 116–19. [Excerpts from an Inaugural Lecture, University of Dundee, given on Wednesday 26 May 1999.]

11. 'Patrick Geddes and Scottish Generalism', in *The City After Patrick Geddes*, edited by Volker Welter and James Lawson (Bern: Peter Lang, 2000), pp. 55–70.

12. 'A European Scot: Patrick Geddes', *Etudes Ecossaises*, No. 6, 2000, pp. 43–6.

13. 'The patron, the professor and the painter: aspects of cultural activity in Dundee at the close of the nineteenth century', in *Victorian Dundee: Image and Realities*, edited by Louise Miskell, Christopher A. Whatley and Bob Harris (East Linton: Tuckwell, 2000), pp. 135–50.

14. 'Anima Celtica: Embodying the Soul of the Nation in 1890s Edinburgh', in *Art, Nation and Gender: Ethnic Landscapes, Myths and Mother-Figures*, edited by Tricia Cusack and Sìghle Bhreathnach-Lynch (Aldershot: Ashgate, 2003), pp. 29–37.

15. 'Patrick Geddes: Environment and Culture', in *Think Global, Act Local: The Life and Legacy of Patrick Geddes*, edited by Walter Stephen (Edinburgh: Luath Press, 2004), pp. 61–84. [Text of the Falkland Stewardship Lecture for 2004, 'Patrick Geddes: Steward of Environment and Culture'].

16. 'Patrick Geddes's Generalism: from Edinburgh's Old Town to Paris's Universal Exhibition', in *Patrick Geddes: the French Connection*, edited by Frances Fowle and Belinda Thomson (Oxford: White Cockade Publishing, 2004), pp. 83–93.

17. 'Patrick Geddes: Science and Art in Dundee', in *The Artist and the Thinker*, edited by Matthew Jarron (Dundee: University of Dundee Museum Services, 2004), pp. 13–29.

18. 'Patrick Geddes and Cultural Renewal through Visual Art: Scotland–India–Japan', in *Patrick Geddes: By Leaves We Live*, edited by Kiyoshi Okutsu, Alan Johnston, Murdo Macdonald and Noboru Sadakata (Yamaguchi: Yamaguchi Institute of Contemporary Art, in association with Edinburgh College of Art, 2005), pp. 46–71.

19. 'Celticism and Internationalism in the Circle of Patrick Geddes', *Visual Culture in Britain*, Vol. 6, No. 2, 2005, pp. 70–83.

20. 'Pointing to Iona: Patrick Geddes, John Duncan and Ananda Coomaraswamy', *Journal of the Scottish Society for Art History*, Vol. 11, 2006, pp. 43–51.

21. 'Macdonald, Murdo, 'The Visual Dimension of *Carmina Gadelica*', in *The Life and Legacy of Alexander Carmichael*, edited by

Domhnall Uilleam Stiùbhart (Port of Ness: Islands Book Trust, 2008), pp. 135–45.

22. 'The Visual Preconditions of Celtic Revival Art in Scotland', *Journal of the Scottish Society for Art History*, Vol. 13, 2009, pp. 16–21.

23. 'Patrick Geddes and the Scottish Generalist Tradition', *Scottish Affairs*, No. 69, Autumn 2009, pp. 40–56. [The Royal Town Planning Institute in Scotland's Sir Patrick Geddes Commemorative Lecture for 2009.]

24. 'Patrick Geddes and the Tradition of Scottish Generalism', *Journal of Scottish Thought*, Vol. 5, 'Patrick Geddes' (Aberdeen: Aberdeen University Press, 2012), pp. 73–87. [An edited and footnoted version of the above.]

25. 'Towards Celtic Revival Art in Scotland: the Visual Background to Revival in the 18th and 19th Centuries', in *Romantic Ireland: From Tone to Gonne*, edited by Paddy Lyons, Willy Maley and John Miller (Newcastle: Cambridge Scholars Publishing, 2013), pp. 340–9.

26. 'Education, Visual Art and Cultural Revival: Tagore, Geddes, Nivedita and Coomaraswamy', *Gitanjali and Beyond*, 1, 2016, pp. 39–57.

27. 'Ossian and Visual Art – Mislaid and Rediscovered', *Journal for Eighteenth Century Studies*, Vol. 39, No. 2, June 2016, pp. 235–48.

28. 'Nivedita from a Scottish Point of View', *Prabuddha Bharata/ Awakened India*, January 2017, pp. 161–7.

Index